Good Laboratory Practice

OECD PRINCIPLES AND GUIDANCE
FOR COMPLIANCE MONITORING

OECD

ORGANISATION FOR ECONOMIC CO-OPERATION AND DEVELOPMENT

ORGANISATION FOR ECONOMIC CO-OPERATION AND DEVELOPMENT

The OECD is a unique forum where the governments of 30 democracies work together to address the economic, social and environmental challenges of globalisation. The OECD is also at the forefront of efforts to understand and to help governments respond to new developments and concerns, such as corporate governance, the information economy and the challenges of an ageing population. The Organisation provides a setting where governments can compare policy experiences, seek answers to common problems, identify good practice and work to co-ordinate domestic and international policies.

The OECD member countries are: Australia, Austria, Belgium, Canada, the Czech Republic, Denmark, Finland, France, Germany, Greece, Hungary, Iceland, Ireland, Italy, Japan, Korea, Luxembourg, Mexico, the Netherlands, New Zealand, Norway, Poland, Portugal, the Slovak Republic, Spain, Sweden, Switzerland, Turkey, the United Kingdom and the United States. The Commission of the European Communities takes part in the work of the OECD.

OECD Publishing disseminates widely the results of the Organisation's statistics gathering and research on economic, social and environmental issues, as well as the conventions, guidelines and standards agreed by its members.

This work is published on the responsibility of the Secretary-General of the OECD. The opinions expressed and arguments employed herein do not necessarily reflect the official views of the Organisation or of the governments of its member countries.

Publié en français sous le titre :
Les bonnes pratiques de laboratoire
PRINCIPES DE L'OCDE ET DIRECTIVES POUR ÉVALUER LEUR RESPECT

© OECD 2005

No reproduction, copy, transmission or translation of this publication may be made without written permission. Applications should be sent to OECD Publishing: *rights@oecd.org* or by fax (33 1) 45 24 13 91. Permission to photocopy a portion of this work should be addressed to the Centre français d'exploitation du droit de copie, 20, rue des Grands-Augustins, 75006 Paris, France (*contact@cfcopies.com*).

Foreword

Chemicals control legislation in OECD member countries is founded in a proactive philosophy of preventing risk by testing and assessing chemicals to determine their potential hazards. The requirement that evaluations of chemicals be based on safety test data of sufficient quality, rigour and reproducibility is a basic principle in this legislation. The OECD Principles of Good Laboratory Practice (GLP) have been developed to promote the quality and validity of test data used for determining the safety of chemicals and chemicals products. They embody a managerial concept covering the organisational processes and the conditions under which non-clinical environmental health and safety studies are planned, performed, monitored, recorded and reported. The Principles of GLP are required to be followed by test facilities carrying out studies to be submitted to national regulatory authorities for the purposes of assessment of chemicals and other uses relating to the protection of man and the environment. The OECD Principles of GLP are an integral part of a legally binding Council Decision on the Mutual Acceptance of Data in the Assessment of Chemicals.

The issue of data quality in the assessment of chemicals has an important international dimension. If regulatory authorities in member countries can rely on safety test data developed abroad, duplicative testing can be avoided and costs saved for both government and industry. Moreover, common principles for GLP for test facilities, and common procedures for governments to monitor compliance with them, facilitate the exchange of information and prevent the emergence of non-tariff barriers to trade, while contributing to the protection of human health and the environment.

The Council Decision-Recommendation on Compliance with GLP [C(89)87(Final)] together with the MAD Council Decisions sets the legal framework for international liaison for this area. This publication gathers together all of the documents published in the OECD Series on GLP and Compliance Monitoring to date – The OECD Principles of GLP, Guidance Documents for Compliance Monitoring Authorities, Consensus Documents and Advisory Documents of the Working Group on GLP – and the three OECD Council Acts related to MAD.

Table of Contents

Part I
The OECD Principles of Good Laboratory Practice

Section I – Introduction 9
Section II – Good Laboratory Practice Principles 13

Part II
Consensus and Advisory Documents on the Application of the Principles of Good Laboratory Practice

Chapter 1. Consensus Documents 27
   1. Quality Assurance and Good Laboratory Practice 28
   2. Compliance of laboratory suppliers with Good Laboratory Practice Principles 35
   3. The application of the Good Laboratory Practice Principles to field studies 39
   4. The application of the Good Laboratory Practice Principles to short-term studies 48
   5. The role and responsibilities of the Study Director in Good Laboratory Pratice studies 56
   6. The application of the Principles of GLP to computerised systems 63
   7. The application of the OECD Principles of GLP to the organisation and management of multi-site studies 73

Chapter 2. Advisory Documents of the Working Group on GLP 83
   8. The role and responsibilities of the sponsor in the application of the Principles of the Laboratory Practice 84
   9. The application of the OECD Principles of GLP to *in vitro* studies 87

Part III
Guidance and Advisory Documents for Good Laboratory Practice Compliance Monitoring Authorities

Chapter 3. Guidance for GLP Monitoring Authorities 101
   1. Guides for Compliance Monitoring Procedures for Good Laboratory Practice 102
   2. Guidance for the conduct of test facility Inspections and study audits 108
   3. Guidance for the preparation of GLP inspection reports 119

Chapter 4. Advisory Document of the Working Group on GLP 123
   1. Requesting and carrying out inspections and study audits
      in another country 124

Annex: OECD Council Acts Related to the Mutual Acceptance of Data............................ 127
   Introduction.. 127
   1. Decision of the Council concerning the mutual acceptance of data
      in the assessment of chemicals [C(81)30(Final)] 128
   2. Council Decision-Recommendation on compliance with Principles
      of Good Laboratory Practice [C(89)87(Final)] 130
   3. Council Decision concerning the adherence of non-member countries to
      the Council Acts related to the mutual acceptance of data in the assessment
      of chemicals [C(81)30(Final) AND C(89)87(Final)] [C(97)114/Final] 137

PART I

The OECD Principles of Good Laboratory Practice*

Government and industry are concerned about the quality of non-clinical health and environmental safety studies upon which hazard assessments are based. As a consequence, OECD member countries have established criteria for the performance of these studies.

To avoid different schemes of implementation that could impede international trade in chemicals, OECD member countries have pursued international harmonisation of test methods and good laboratory practice. In 1979 and 1980, an international group of experts established under the Special Programme on the Control of Chemicals developed the OECD Principles of Good Laboratory Practice (GLP), utilising common managerial and scientific practices and experience from various national and international sources. These Principles of GLP were adopted by the OECD Council in 1981, as an Annex to the Council Decision on the Mutual Acceptance of Data in the Assessment of Chemicals [C(81)30(Final)].

In 1995 and 1996, a new group of experts was formed to revise and update the Principles. The current document is the result of the consensus reached by that group. It cancels and replaces the original Principles adopted in 1981.

The purpose of these Principles of Good Laboratory Practice is to promote the development of quality test data. Comparable quality of test data forms the basis for the mutual acceptance of data among countries. If individual countries can confidently rely on test data developed in other countries, duplicative testing can be avoided, thereby saving time and resources. The application of these Principles should help to avoid the creation of technical barriers to trade, and further improve the protection of human health and the environment.

---

* No. 1 in the OECD Series on Principles of Good Laboratory Practice and Compliance Monitoring.

SECTION I – INTRODUCTION

History

The OECD Principles of Good Laboratory Practice were first developed by an Expert Group on GLP established in 1978 under the Special Programme on the Control of Chemicals. The GLP regulations for non-clinical laboratory studies published by the US Food and Drug Administration in 1976 provided the basis for the work of the Expert Group, which was led by the United States and comprised experts from the following countries and organisations: Australia, Austria, Belgium, Canada, Denmark, France, the Federal Republic of Germany, Greece, Italy, Japan, the Netherlands, New Zealand, Norway, Sweden, Switzerland, the United Kingdom, the United States, the Commission of the European Communities, the World Health Organisation and the International Organisation for Standardisation.

Those Principles of GLP were formally recommended for use in member countries by the OECD Council in 1981. They were set out (in Annex II) as an integral part of the Council Decision on Mutual Acceptance of Data in the Assessment of Chemicals, which states that "data generated in the testing of chemicals in an OECD member country in accordance with OECD Test Guidelines (OECD Guidelines for the Testing of Chemicals, 1981 and continuing series) and OECD Principles of Good Laboratory Practice shall be accepted in other member countries for purposes of assessment and other uses relating to the protection of man and the environment" [C(81)30(Final)] (see Annex, page 130).

After a decade and half of use, member countries considered that there was a need to review and update the Principles of GLP to account for scientific and technical progress in the field of safety testing and the fact that safety testing was currently required in many more areas than was the case at the end of the 1970s. On the proposal of the Joint Meeting of the Chemicals Group and Management Committee of the Special Programme on the Control of Chemicals, another Expert Group was therefore established in 1995 to develop a proposal to revise the Principles of GLP. The Expert Group, which completed its work in 1996, was led by Germany and comprised experts from Australia, Austria, Belgium, Canada, the Czech Republic, Denmark, Finland, France, Germany, Greece, Hungary, Ireland, Italy, Japan, Korea, the Netherlands, Norway, Poland, Portugal, the Slovak Republic, Spain, Sweden, Switzerland, the United Kingdom, the United States and the International Organisation for Standardisation.

The Revised OECD Principles of GLP were reviewed in the relevant policy bodies of the Organisation and were adopted by Council on 26 November, 1997 [C(97)186/Final], which formally amended Annex II of the 1981 Council Decision.

1. Scope

These Principles of Good Laboratory Practice should be applied to the non-clinical safety testing of test items contained in pharmaceutical products, pesticide products, cosmetic products, veterinary drugs as well as food additives, feed additives, and industrial chemicals. These test items are frequently synthetic chemicals, but may be of natural or biological origin and, in some circumstances, may be living organisms. The purpose of testing these test items is to obtain data on their properties and/or their safety with respect to human health and/or the environment.

Non-clinical health and environmental safety studies covered by the Principles of Good Laboratory Practice include work conducted in the laboratory, in greenhouses, and in the field.

Unless specifically exempted by national legislation, these Principles of Good Laboratory Practice apply to all non-clinical health and environmental safety studies required by regulations for the purpose of registering or licensing pharmaceuticals, pesticides, food and feed additives, cosmetic products, veterinary drug products and similar products, and for the regulation of industrial chemicals.

2. Definitions of terms

2.1. Good Laboratory Practice

1. *Good Laboratory Practice* (GLP) is a quality system concerned with the organisational process and the conditions under which non-clinical health and environmental safety studies are planned, performed, monitored, recorded, archived and reported.

2.2. Terms concerning the organisation of a test facility

1. *Test facility* means the persons, premises and operational unit(s) that are necessary for conducting the non-clinical health and environmental safety study. For multi-site studies, those which are conducted at more than one site, the test facility comprises the site at which the Study Director is located and all individual test sites, which individually or collectively can be considered to be test facilities.

2. *Test site* means the location(s) at which a phase(s) of a study is conducted.

3. *Test facility management* means the person(s) who has the authority and formal responsibility for the organisation and functioning of the test facility according to these Principles of Good Laboratory Practice.

4. *Test site management* (if appointed) means the person(s) responsible for ensuring that the phase(s) of the study, for which he is responsible, are conducted according to these Principles of Good Laboratory Practice.

5. *Sponsor* means an entity which commissions, supports and/or submits a non-clinical health and environmental safety study.

6. *Study Director* means the individual responsible for the overall conduct of the non-clinical health and environmental safety study.

7. *Principal Investigator* means an individual who, for a multi-site study, acts on behalf of the Study Director and has defined responsibility for delegated phases of the study. The Study Director's responsibility for the overall conduct of the study cannot be delegated to the Principal Investigator(s); this includes approval of the study plan and its amendments, approval of the final report, and ensuring that all applicable Principles of Good Laboratory Practice are followed.

8. *Quality Assurance Programme* means a defined system, including personnel, which is independent of study conduct and is designed to assure test facility management of compliance with these Principles of Good Laboratory Practice.

9. *Standard Operating Procedures (SOPs)* means documented procedures which describe how to perform tests or activities normally not specified in detail in study plans or test guidelines.

10. *Master schedule* means a compilation of information to assist in the assessment of workload and for the tracking of studies at a test facility.

2.3. Terms concerning the non-clinical health and environmental safety study

1. *Non-clinical health and environmental safety study*, henceforth referred to simply as "study", means an experiment or set of experiments in which a test item is examined under laboratory conditions or in the environment to obtain data on its properties and/or its safety, intended for submission to appropriate regulatory authorities.

2. *Short-term study* means a study of short duration with widely used, routine techniques.

3. *Study plan* means a document which defines the objectives and experimental design for the conduct of the study, and includes any amendments.

4. *Study plan amendment* means an intended change to the study plan after the study initiation date.

5. *Study plan deviation* means an unintended departure from the study plan after the study initiation date.

6. *Test system* means any biological, chemical or physical system or a combination thereof used in a study.

7. *Raw data* means all original test facility records and documentation, or verified copies thereof, which are the result of the original observations and activities in a study. Raw data also may include, for example, photographs, microfilm or microfiche copies, computer readable media, dictated observations, recorded data from automated instruments, or any other data storage medium that has been recognised as capable of providing secure storage of information for a time period as stated in Section 10, below.

8. *Specimen* means any material derived from a test system for examination, analysis, or retention.

9. *Experimental starting date* means the date on which the first study specific data are collected.

10. *Experimental completion date* means the last date on which data are collected from the study.

11. *Study initiation date* means the date the Study Director signs the study plan.

12. *Study completion date* means the date the Study Director signs the final report.

2.4. Terms concerning the test item

1. *Test item* means an article that is the subject of a study.

2. *Reference item* ("control item") means any article used to provide a basis for comparison with the test item.

3. *Batch* means a specific quantity or lot of a test item or reference item produced during a defined cycle of manufacture in such a way that it could be expected to be of a uniform character and should be designated as such.

4. *Vehicle* means any agent which serves as a carrier used to mix, disperse, or solubilise the test item or reference item to facilitate the administration/application to the test system.

SECTION II – GOOD LABORATORY PRACTICE PRINCIPLES

1. Test facility organisation and personnel

1.1. *Test facility management's responsibilities*

1. Each test facility management should ensure that these Principles of Good Laboratory Practice are complied with, in its test facility.

2. At a minimum it should:

    a) Ensure that a statement exists which identifies the individual(s) within a test facility who fulfil the responsibilities of management as defined by these Principles of Good Laboratory Practice.

    b) Ensure that a sufficient number of qualified personnel, appropriate facilities, equipment, and materials are available for the timely and proper conduct of the study.

    c) Ensure the maintenance of a record of the qualifications, training, experience and job description for each professional and technical individual.

    d) Ensure that personnel clearly understand the functions they are to perform and, where necessary, provide training for these functions.

    e) Ensure that appropriate and technically valid Standard Operating Procedures are established and followed, and approve all original and revised Standard Operating Procedures.

    f) Ensure that there is a Quality Assurance Programme with designated personnel and assure that the quality assurance responsibility is being performed in accordance with these Principles of Good Laboratory Practice.

    g) Ensure that for each study an individual with the appropriate qualifications, training, and experience is designated by the management as the Study Director before the study is initiated. Replacement of a Study Director should be done according to established procedures, and should be documented.

    h) Ensure, in the event of a multi-site study, that, if needed, a Principal Investigator is designated, who is appropriately trained, qualified and experienced to supervise the delegated phase(s) of the study. Replacement of a Principal Investigator should be done according to established procedures, and should be documented.

    i) Ensure documented approval of the study plan by the Study Director.

    j) Ensure that the Study Director has made the approved study plan available to the Quality Assurance personnel.

    k) Ensure the maintenance of a historical file of all Standard Operating Procedures.

    l) Ensure that an individual is identified as responsible for the management of the archive(s).

m) Ensure the maintenance of a master schedule.

n) Ensure that test facility supplies meet requirements appropriate to their use in a study.

o) Ensure for a multi-site study that clear lines of communication exist between the Study Director, Principal Investigator(s), the Quality Assurance Programme(s) and study personnel.

p) Ensure that test and reference items are appropriately characterised.

q) Establish procedures to ensure that computerised systems are suitable for their intended purpose, and are validated, operated and maintained in accordance with these Principles of Good Laboratory Practice.

3. When a phase(s) of a study is conducted at a test site, test site management (if appointed) will have the responsibilities as defined above with the following exceptions: 1.1.2g), i), j) and o).

1.2. Study Director's Responsibilities

1. The Study Director is the single point of study control and has the responsibility for the overall conduct of the study and for its final report.

2. These responsibilities should include, but not be limited to, the following functions. The Study Director should:

   a) Approve the study plan and any amendments to the study plan by dated signature.

   b) Ensure that the Quality Assurance personnel have a copy of the study plan and any amendments in a timely manner and communicate effectively with the Quality Assurance personnel as required during the conduct of the study.

   c) Ensure that study plans and amendments and Standard Operating Procedures are available to study personnel.

   d) Ensure that the study plan and the final report for a multi-site study identify and define the role of any Principal Investigator(s) and any test facilities and test sites involved in the conduct of the study.

   e) Ensure that the procedures specified in the study plan are followed, and assess and document the impact of any deviations from the study plan on the quality and integrity of the study, and take appropriate corrective action if necessary; acknowledge deviations from Standard Operating Procedures during the conduct of the study.

   f) Ensure that all raw data generated are fully documented and recorded.

   g) Ensure that computerised systems used in the study have been validated.

   h) Sign and date the final report to indicate acceptance of responsibility for the validity of the data and to indicate the extent to which the study complies with these Principles of Good Laboratory Practice.

   i) Ensure that after completion (including termination) of the study, the study plan, the final report, raw data and supporting material are archived.

1.3. Principal Investigator's responsibilities

The Principal Investigator will ensure that the delegated phases of the study are conducted in accordance with the applicable Principles of Good Laboratory Practice.

1.4. Study personnel's responsibilities

1. All personnel involved in the conduct of the study must be knowledgeable in those parts of the Principles of Good Laboratory Practice which are applicable to their involvement in the study.

2. Study personnel will have access to the study plan and appropriate Standard Operating Procedures applicable to their involvement in the study. It is their responsibility to comply with the instructions given in these documents. Any deviation from these instructions should be documented and communicated directly to the Study Director, and/or if appropriate, the Principal Investigator(s).

3. All study personnel are responsible for recording raw data promptly and accurately and in compliance with these Principles of Good Laboratory Practice, and are responsible for the quality of their data.

4. Study personnel should exercise health precautions to minimise risk to themselves and to ensure the integrity of the study. They should communicate to the appropriate person any relevant known health or medical condition in order that they can be excluded from operations that may affect the study.

2. Quality Assurance Programme

2.1. General

1. The test facility should have a documented Quality Assurance Programme to assure that studies performed are in compliance with these Principles of Good Laboratory Practice.

2. The Quality Assurance Programme should be carried out by an individual or by individuals designated by and directly responsible to management and who are familiar with the test procedures.

3. This individual(s) should not be involved in the conduct of the study being assured.

2.2. Responsibilities of the Quality Assurance personnel

1. The responsibilities of the Quality Assurance personnel include, but are not limited to, the following functions. They should:

    a) Maintain copies of all approved study plans and Standard Operating Procedures in use in the test facility and have access to an up-to-date copy of the master schedule.

    b) Verify that the study plan contains the information required for compliance with these Principles of Good Laboratory Practice. This verification should be documented.

    c) Conduct inspections to determine if all studies are conducted in accordance with these Principles of Good Laboratory Practice. Inspections should also determine that study plans and Standard Operating Procedures have been made available to study personnel and are being followed.

Inspections can be of three types as specified by Quality Assurance Programme Standard Operating Procedures:

- study-based inspections;
- facility-based inspections;
- process-based inspections.

Records of such inspections should be retained.

d) Inspect the final reports to confirm that the methods, procedures, and observations are accurately and completely described, and that the reported results accurately and completely reflect the raw data of the studies.

e) Promptly report any inspection results in writing to management and to the Study Director, and to the Principal Investigator(s) and the respective management, when applicable.

f) Prepare and sign a statement, to be included with the final report, which specifies types of inspections and their dates, including the phase(s) of the study inspected, and the dates inspection results were reported to management and the Study Director and Principal Investigator(s), if applicable. This statement would also serve to confirm that the final report reflects the raw data.

3. Facilities

3.1. General

1. The test facility should be of suitable size, construction and location to meet the requirements of the study and to minimise disturbance that would interfere with the validity of the study.
2. The design of the test facility should provide an adequate degree of separation of the different activities to assure the proper conduct of each study.

3.2. Test system facilities

1. The test facility should have a sufficient number of rooms or areas to assure the isolation of test systems and the isolation of individual projects, involving substances or organisms known to be or suspected of being biohazardous.
2. Suitable rooms or areas should be available for the diagnosis, treatment and control of diseases, in order to ensure that there is no unacceptable degree of deterioration of test systems.
3. There should be storage rooms or areas as needed for supplies and equipment. Storage rooms or areas should be separated from rooms or areas housing the test systems and should provide adequate protection against infestation, contamination, and/or deterioration.

3.3. Facilities for handling test and reference items

1. To prevent contamination or mix-ups, there should be separate rooms or areas for receipt and storage of the test and reference items, and mixing of the test items with a vehicle.

2. Storage rooms or areas for the test items should be separate from rooms or areas containing the test systems. They should be adequate to preserve identity, concentration, purity, and stability, and ensure safe storage for hazardous substances.

3.4. Archive facilities

Archive facilities should be provided for the secure storage and retrieval of study plans, raw data, final reports, samples of test items and specimens. Archive design and archive conditions should protect contents from untimely deterioration.

3.5. Waste disposal

Handling and disposal of wastes should be carried out in such a way as not to jeopardise the integrity of studies. This includes provision for appropriate collection, storage and disposal facilities, and decontamination and transportation procedures.

4. Apparatus, material and reagents

1. Apparatus, including validated computerised systems, used for the generation, storage and retrieval of data, and for controlling environmental factors relevant to the study should be suitably located and of appropriate design and adequate capacity.
2. Apparatus used in a study should be periodically inspected, cleaned, maintained, and calibrated according to Standard Operating Procedures. Records of these activities should be maintained. Calibration should, where appropriate, be traceable to national or international standards of measurement.
3. Apparatus and materials used in a study should not interfere adversely with the test systems.
4. Chemicals, reagents, and solutions should be labelled to indicate identity (with concentration if appropriate), expiry date and specific storage instructions. Information concerning source, preparation date and stability should be available. The expiry date may be extended on the basis of documented evaluation or analysis.

5. Test systems

5.1. Physical/chemical

1. Apparatus used for the generation of physical/chemical data should be suitably located and of appropriate design and adequate capacity.
2. The integrity of the physical/chemical test systems should be ensured.

5.2. Biological

1. Proper conditions should be established and maintained for the storage, housing, handling and care of biological test systems, in order to ensure the quality of the data.
2. Newly received animal and plant test systems should be isolated until their health status has been evaluated. If any unusual mortality or morbidity occurs, this lot should not be used in studies and, when appropriate, should be humanely destroyed. At the experimental starting date of a study, test systems should be free of any disease or condition that might interfere with the purpose or conduct of the study. Test systems that become diseased or injured during the course of a study should be

isolated and treated, if necessary to maintain the integrity of the study. Any diagnosis and treatment of any disease before or during a study should be recorded.

3. Records of source, date of arrival, and arrival condition of test systems should be maintained.

4. Biological test systems should be acclimatised to the test environment for an adequate period before the first administration/application of the test or reference item.

5. All information needed to properly identify the test systems should appear on their housing or containers. Individual test systems that are to be removed from their housing or containers during the conduct of the study should bear appropriate identification, wherever possible.

6. During use, housing or containers for test systems should be cleaned and sanitised at appropriate intervals. Any material that comes into contact with the test system should be free of contaminants at levels that would interfere with the study. Bedding for animals should be changed as required by sound husbandry practice. Use of pest control agents should be documented.

7. Test systems used in field studies should be located so as to avoid interference in the study from spray drift and from past usage of pesticides.

6. Test and reference items

6.1. Receipt, handling, sampling and storage

1. Records including test item and reference item characterisation, date of receipt, expiry date, quantities received and used in studies should be maintained.

2. Handling, sampling, and storage procedures should be identified in order that the homogeneity and stability are assured to the degree possible and contamination or mix-up are precluded.

3. Storage container(s) should carry identification information, expiry date, and specific storage instructions.

6.2. Characterisation

1. Each test and reference item should be appropriately identified (*e.g.* code, Chemical Abstracts Service Registry Number [CAS number], name, biological parameters).

2. For each study, the identity, including batch number, purity, composition, concentrations, or other characteristics to appropriately define each batch of the test or reference items should be known.

3. In cases where the test item is supplied by the sponsor, there should be a mechanism, developed in co-operation between the sponsor and the test facility, to verify the identity of the test item subject to the study.

4. The stability of test and reference items under storage and test conditions should be known for all studies.

5. If the test item is administered or applied in a vehicle, the homogeneity, concentration and stability of the test item in that vehicle should be determined. For test items used in field studies (*e.g.* tank mixes), these may be determined through separate laboratory experiments.

6. A sample for analytical purposes from each batch of test item should be retained for all studies except short-term studies.

7. Standard Operating Procedures

7.1. A test facility should have written Standard Operating Procedures approved by test facility management that are intended to ensure the quality and integrity of the data generated by that test facility. Revisions to Standard Operating Procedures should be approved by test facility management.

7.2. Each separate test facility unit or area should have immediately available current Standard Operating Procedures relevant to the activities being performed therein. Published text books, analytical methods, articles and manuals may be used as supplements to these Standard Operating Procedures.

7.3. Deviations from Standard Operating Procedures related to the study should be documented and should be acknowledged by the Study Director and the Principal Investigator(s), as applicable.

7.4. Standard Operating Procedures should be available for, but not be limited to, the following categories of test facility activities. The details given under each heading are to be considered as illustrative examples.

1. *Test and reference items*
   Receipt, identification, labelling, handling, sampling and storage.

2. *Apparatus, materials and reagents*
   a) *Apparatus*
   Use, maintenance, cleaning and calibration.

   b) *Computerised systems*
   Validation, operation, maintenance, security, change control and back-up.

   c) *Materials, reagents and solutions*
   Preparation and labelling.

3. *Record keeping, reporting, storage, and retrieval*
   Coding of studies, data collection, preparation of reports, indexing systems, handling of data, including the use of computerised systems.

4. *Test system (where appropriate)*
   a) Room preparation and environmental room conditions for the test system.

   b) Procedures for receipt, transfer, proper placement, characterisation, identification and care of the test system.

   c) Test system preparation, observations and examinations, before, during and at the conclusion of the study.

   d) Handling of test system individuals found moribund or dead during the study.

   e) Collection, identification and handling of specimens including necropsy and histopathology.

   f) Siting and placement of test systems in test plots.

5. *Quality Assurance Procedures*
   Operation of Quality Assurance personnel in planning, scheduling, performing, documenting and reporting inspections.

8. Performance of the study

8.1. Study plan

1. For each study, a written plan should exist prior to the initiation of the study. The study plan should be approved by dated signature of the Study Director and verified for GLP compliance by Quality Assurance personnel as specified in Section 2.2.1.*b*), above. The study plan should also be approved by the test facility management and the sponsor, if required by national regulation or legislation in the country where the study is being performed.

2. a) Amendments to the study plan should be justified and approved by dated signature of the Study Director and maintained with the study plan.

    b) Deviations from the study plan should be described, explained, acknowledged and dated in a timely fashion by the Study Director and/or Principal Investigator(s) and maintained with the study raw data.

3. For short-term studies, a general study plan accompanied by a study specific supplement may be used.

8.2. Content of the study plan

The study plan should contain, but not be limited to the following information:

1. *Identification of the study, the test item and reference item*
    a) A descriptive title.
    b) A statement which reveals the nature and purpose of the study.
    c) Identification of the test item by code or name (IUPAC; CAS number, biological parameters, etc.).
    d) The reference item to be used.

2. *Information concerning the sponsor and the test facility*
    a) Name and address of the sponsor.
    b) Name and address of any test facilities and test sites involved.
    c) Name and address of the Study Director.
    d) Name and address of the Principal Investigator(s), and the phase(s) of the study delegated by the Study Director and under the responsibility of the Principal Investigator(s).

3. *Dates*
    a) The date of approval of the study plan by signature of the Study Director. The date of approval of the study plan by signature of the test facility management and sponsor, if required by national regulation or legislation in the country where the study is being performed.
    b) The proposed experimental starting and completion dates.

4. *Test methods*

    Reference to the OECD Test Guideline or other test guideline or method to be used.

5. *Issues (where applicable)*
    a) The justification for selection of the test system.

- b) Characterisation of the test system, such as the species, strain, substrain, source of supply, number, body weight range, sex, age and other pertinent information.
- c) The method of administration and the reason for its choice.
- d) The dose levels and/or concentration(s), frequency, and duration of administration/application.
- e) Detailed information on the experimental design, including a description of the chronological procedure of the study, all methods, materials and conditions, type and frequency of analysis, measurements, observations and examinations to be performed, and statistical methods to be used (if any).

6. *Records*

   A list of records to be retained.

8.3. Conduct of the study

1. A unique identification should be given to each study. All items concerning this study should carry this identification. Specimens from the study should be identified to confirm their origin. Such identification should enable traceability, as appropriate for the specimen and study.
2. The study should be conducted in accordance with the study plan.
3. All data generated during the conduct of the study should be recorded directly, promptly, accurately, and legibly by the individual entering the data. These entries should be signed or initialled and dated.
4. Any change in the raw data should be made so as not to obscure the previous entry, should indicate the reason for change and should be dated and signed or initialled by the individual making the change.
5. Data generated as a direct computer input should be identified at the time of data input by the individual(s) responsible for direct data entries. Computerised system design should always provide for the retention of full audit trails to show all changes to the data without obscuring the original data. It should be possible to associate all changes to data with the persons having made those changes, for example, by use of timed and dated (electronic) signatures. Reason for changes should be given.

9. Reporting of study results

9.1. General

1. A final report should be prepared for each study. In the case of short term studies, a standardised final report accompanied by a study specific extension may be prepared.
2. Reports of Principal Investigators or scientists involved in the study should be signed and dated by them.
3. The final report should be signed and dated by the Study Director to indicate acceptance of responsibility for the validity of the data. The extent of compliance with these Principles of Good Laboratory Practice should be indicated.
4. Corrections and additions to a final report should be in the form of amendments. Amendments should clearly specify the reason for the corrections or additions and should be signed and dated by the Study Director.

5. Reformatting of the final report to comply with the submission requirements of a national registration or regulatory authority does not constitute a correction, addition or amendment to the final report.

9.2. *Content of the final report*

The final report should include, but not be limited to, the following information:

1. *Identification of the study, the test item and reference item*
    a) A descriptive title.
    b) Identification of the test item by code or name (IUPAC, CAS number, biological parameters, etc.).
    c) Identification of the reference item by name.
    d) Characterisation of the test item including purity, stability and homogeneity.

2. *Information concerning the sponsor and the test facility*
    a) Name and address of the sponsor.
    b) Name and address of any test facilities and test sites involved.
    c) Name and address of the Study Director.
    d) Name and address of the Principal Investigator(s) and the phase(s) of the study delegated, if applicable.
    e) Name and address of scientists having contributed reports to the final report.

3. *Dates*

    Experimental starting and completion dates.

4. *Statement*

    A Quality Assurance Programme statement listing the types of inspections made and their dates, including the phase(s) inspected, and the dates any inspection results were reported to management and to the Study Director and Principal Investigator(s), if applicable. This statement would also serve to confirm that the final report reflects the raw data.

5. *Description of materials and test methods*
    a) Description of methods and materials used.
    b) Reference to OECD Test Guideline or other test guideline or method.

6. *Results*
    a) A summary of results.
    b) All information and data required by the study plan.
    c) A presentation of the results, including calculations and determinations of statistical significance.
    d) An evaluation and discussion of the results and, where appropriate, conclusions.

7. *Storage*

    The location(s) where the study plan, samples of test and reference items, specimens, raw data and the final report are to be stored.

10. Storage and retention of records and materials

10.1. The following should be retained in the archives for the period specified by the appropriate authorities:

a) The study plan, raw data, samples of test and reference items, specimens, and the final report of each study.

b) Records of all inspections performed by the Quality Assurance Programme, as well as master schedules.

c) Records of qualifications, training, experience and job descriptions of personnel.

d) Records and reports of the maintenance and calibration of apparatus.

e) Validation documentation for computerised systems.

f) The historical file of all Standard Operating Procedures.

g) Environmental monitoring records.

In the absence of a required retention period, the final disposition of any study materials should be documented. When samples of test and reference items and specimens are disposed of before the expiry of the required retention period for any reason, this should be justified and documented. Samples of test and reference items and specimens should be retained only as long as the quality of the preparation permits evaluation.

10.2. Material retained in the archives should be indexed so as to facilitate orderly storage and retrieval.

10.3. Only personnel authorised by management should have access to the archives. Movement of material in and out of the archives should be properly recorded.

10.4. If a test facility or an archive contracting facility goes out of business and has no legal successor, the archive should be transferred to the archives of the sponsor(s) of the study(ies).

PART II

Consensus and Advisory Documents on the Application of the Principles of Good Laboratory Practice

The seven Consensus Documents reproduced here were developed by Consensus Workshops comprising representatives of member countries (from GLP compliance monitoring authorities and national data receiving authorities) and other stakeholders (from test facilities and relevant international organisations). The two Advisory Documents were developed by the Working Group on GLP, with the assistance of experts, as appropriate. These documents have all been endorsed by the Working Group on GLP* and by the Joint Meeting of the Chemicals Committee and Working Party on Chemicals, Pesticides and Biotechnology, and declassified under the authority of the Secretary-General. Their purpose is to assist governments and test facilities to interpret and apply the OECD Principles of GLP.

* This body, which comprises the Heads of GLP compliance monitoring authorities in OECD and adhering non-member countries, was formerly called "Panel on GLP".

PART II

Chapter 1

Consensus Documents

1. QUALITY ASSURANCE AND GOOD LABORATORY PRACTICE*

History

In the framework of the OECD Consensus Workshop on Good Laboratory Practice, held 16-18 October 1990 in Bad Dürkheim, Germany, a working group met to discuss and arrive at consensus on Good Laboratory Practice and the role of Quality Assurance (QA). The working group was chaired by Dr. Hans Könemann (Head, GLP Compliance Monitoring Authority, the Netherlands). Participants were mainly members of national GLP compliance monitoring units or experienced QA managers from test facilities. The following countries were represented: Austria, Belgium, France, Germany, Ireland, the Netherlands, Norway, Spain, Sweden, Switzerland, and the United Kingdom.

The working group reached consensus on the role of QA as an important component of GLP. It identified major issues related to QA and GLP, but did not attempt to treat the subject exhaustively. One area not specifically addressed was the application of QA to field studies. This and some other aspects of QA will be addressed separately.

The draft consensus document developed by the working group was circulated to member countries and revised, based on the comments received. It was subsequently endorsed by the OECD Panel on GLP, and the Chemicals Group and Management Committee of the Special Programme on the Control of Chemicals. The Environment Committee then recommended that this document be derestricted under the authority of the Secretary-General.

In light of the adoption of the Revised OECD Principles of GLP in 1997, this Consensus Document was reviewed by the Working Group on GLP and revised to make it consistent with modifications made to the Principles. It was endorsed by the Working Group in April 1999 and, subsequently by the Joint Meeting of the Chemicals Committee and Working Party on Chemicals, Pesticides and Biotechnology in August 1999. It is declassified under the authority of the Secretary-General.

Introduction

The OECD Principles of GLP have been in force for over fifteen years. Valuable experience has been gained at test facilities where these principles have been applied, as well as by governmental bodies monitoring for compliance. In light of this experience, some additional guidance can be given on the role and operation of Quality Assurance programmes in test facilities.

---

* No. 11 in the OECD Series on Principles of Good Laboratory Practice and Compliance Monitoring.

References to Quality Assurance in the OECD Principles of Good Laboratory Practice

A Quality Assurance Programme is defined in the Revised OECD Principles of Good Laboratory Practice as "a defined system, including personnel, which is independent of study conduct and is designed to assure test facility management of compliance with these Principles of Good Laboratory Practice" [Section I.2.2.8)]. The responsibilities of the management of a test facility include ensuring "that there is a Quality Assurance Programme with designated personnel and assure that the quality assurance responsibility is being performed in compliance with these Principles of Good Laboratory Practice" [Section II.1.1.2f)]. In addition, the test facility management should ensure "that the Study Director has made the approved study plan available to the Quality Assurance personnel" [Section II.1.1.2j)] and the responsibility of the Study Director should include ensuring "that the Quality Assurance personnel have a copy of the study plan and any amendments in a timely manner and communicate effectively with the Quality Assurance personnel as required during the conduct of the study" [Section II.1.2.2b)]. The test facility management should also ensure that "for a multi-site study that clear lines" of communication exist between the Study Director, Principal Investigator(s), the Quality Assurance Programme(s) and study personnel [Section II.1.1.2o)].

In Section II.2 ("Quality Assurance Programme") the following requirements are listed:

1.1. General

1. The test facility should have a documented Quality Assurance Programme to assure that studies performed are in compliance with these Principles of Good Laboratory Practice.

2. The Quality Assurance Programme should be carried out by an individual or by individuals designated by and directly responsible to management and who are familiar with the test procedures.

3. This individual(s) should not be involved in the conduct of the study being assured.

1.2. Responsibilities of the Quality Assurance personnel

1. The responsibilities of the Quality Assurance personnel include, but are not limited to, the following functions. They should:

    a) Maintain copies of all approved study plans and Standard Operating Procedures in use in the test facility and have access to an up-to-date copy of the master schedule.

    b) Verify that the study plan contains the information required for compliance with these Principles of Good Laboratory Practice. This verification should be documented.

    c) Conduct inspections to determine if all studies are conducted in compliance with these Principles of Good Laboratory Practice. Inspections should also determine that study plans and Standard Operating Procedures have been made available to study personnel and are being followed.

    Inspections can be of three types as specified by Quality Assurance Programme Standard Operating Procedures:

    – study-based inspections;

    – facility-based inspections;

– process-based inspections.

Records of such inspections should be retained.

d) Inspect the final reports to confirm that the methods, procedures, and observations are accurately and completely described, and that the reported results accurately and completely reflect the raw data of the studies.

e) Promptly report any inspection results in writing to management and to the Study Director, and to the Principal Investigator(s) and the respective management, when applicable.

f) Prepare and sign a statement, to be included with the final report, which specifies types of inspections and their dates, including the phase(s) of the study inspected, and the dates inspection results were reported to management and the Study Director and Principal Investigator(s), if applicable. This statement would also serve to confirm that the final report reflects the raw data.

In Section II.7.4.5, the "operation of Quality Assurance personnel in planning, scheduling, performing, documenting and reporting inspections" is one of the categories of laboratory activities for which Standard Operating Procedures (SOPs) should be available.

In Section II.9.2.4, a final study report is required to include "a Quality Assurance Programme statement listing the types of inspections made and their dates, including the phase(s) inspected, and the dates any inspection results were reported to management and to the Study Director and Principal Investigator(s), if applicable. This statement would also serve to confirm that the final report reflects the raw data".

Finally, in Section II.10.1.b), "records of all inspections performed by the Quality Assurance Programme, as well as master schedules" should be retained in the archives for the period specified by the appropriate authorities.

The Quality Assurance-management link

Management of a test facility has the ultimate responsibility for ensuring that the facility as a whole operates in compliance with GLP Principles. Management may delegate designated control activities through the line management organisation, but always retains overall responsibility. An essential management responsibility is the appointment and effective organisation of an adequate number of appropriately qualified and experienced staff throughout the facility, including those specifically required to perform QA functions.

The manager ultimately responsible for GLP should be clearly identified. This person's responsibilities include the appointment of appropriately qualified personnel for both the experimental programme and for the conduct of an independent QA function. Delegation to QA of tasks which are attributed to management in the GLP Principles must not compromise the independence of the QA operation, and must not entail any involvement of QA personnel in the conduct of the study other than in a monitoring role. The person appointed to be responsible for QA must have direct access to the different levels of management, particularly to top level management of the test facility.

Qualifications of Quality Assurance personnel

Quality Assurance personnel should have the training, expertise and experience necessary to fulfil their responsibilities. They must be familiar with the test procedures, standards and systems operated at or on behalf of the test facility.

Individuals appointed to QA functions should have the ability to understand the basic concepts underlying the activities being monitored. They should also have a thorough understanding of the Principles of GLP.

In case of lack of specialised knowledge, or the need for a second opinion, it is recommended that the QA operation ask for specialist support. Management should also ensure that there is a documented training programme encompassing all aspects of QA work. The training programme should, where possible, include on-the-job experience under the supervision of competent and trained staff. Attendance at in-house and external seminars and courses may also be relevant. For example, training in communication techniques and conflict handling is advisable. Training should be continuous and subject to periodic review.

The training of QA personnel must be documented and their competence evaluated. These records should be kept up-to-date and be retained.

Quality Assurance Involvement in developing Standard Operating Procedures and study plans

Management is responsible for ensuring that Standard Operating Procedures (SOPs) are produced, issued, distributed and retained. QA personnel are not normally involved in drafting SOPs; however it is desirable that they review SOPs before use in order to assess their clarity and compliance with GLP Principles.

Management should ensure that the study plan is available to QA before the experimental starting date of the study. This allows QA:

- to monitor compliance of the study plan with GLP;
- to assess the clarity and consistency of the study plan;
- to identify the critical phases of the study; and
- to plan a monitoring programme in relation to the study.

As and when amendments are made to the study plan, they should be copied to QA to facilitate effective study monitoring.

Quality Assurance inspections

QA programmes are frequently based upon the following types of inspections:

- *Study-based inspections*: These are scheduled according to the chronology of a given study, usually by first identifying the critical phases of the study.
- *Facility-based inspections*: These are not based upon specific studies, but cover the general facilities and activities within a laboratory (installations, support services, computer system, training, environmental monitoring, maintenance, calibration, etc.).
- *Process-based inspections*: Again, these are performed independently of specific studies. They are conducted to monitor procedures or processes of a repetitive nature and are generally performed on a random basis. These inspections take place when a process is undertaken very frequently within a laboratory and it is therefore considered inefficient or impractical to undertake study-based inspections. It is recognised that performance of process-based inspections covering phases which occur with a very high frequency may result in some studies not being inspected on an individual basis during their experimental phases.

Quality Assurance planning and justification of Quality Assurance activities and methods

QA should plan its work properly and its planning procedures as well as the operation of QA personnel in performing, documenting and reporting inspections should be described in SOPs. A list of studies planned and in progress should be kept. QA should have access to an up-to-date copy of the master schedule. Such a list is necessary for planning QA activities and assessing the QA workload in the laboratory.

As is the case for any other operative procedures covered by the GLP Principles, the QA programme of inspections and audits should be subject to management verification. Both the QA staff and management should be able to justify the methods chosen for the performance of their tasks.

Quality Assurance inspection reports

National GLP monitoring authorities may request information relating to the types of inspections and their dates, including the phase(s) of the study inspected. However, QA inspection reports should not normally be examined for their contents by national monitoring authorities as this may inhibit QA when preparing inspection reports. Nevertheless, national monitoring authorities may occasionally require access to the contents of inspection reports in order to verify the adequate functioning of QA. They should not inspect such reports merely as an easy way to identify inadequacies in the studies carried out.

Audits of data and final reports

In the GLP Principles, raw data are defined as "all original test facility records and documentation, or verified copies thereof, which are the result of the original observations and activities in a study". Raw data also may include, for example, photographs, microfilm or microfiche copies, computer readable media, dictated observations, recorded data from automated instruments, or any other data storage medium, that has been recognised as capable of providing secure storage of information for a time period as stated in Section 10 below (Section I.2.3.7).

The review of a study's raw data by QA can be carried out in a number of ways. For example, the records may be examined by QA during experimental phases of the study, during process inspections or during audits of final reports. Management should ensure that all final reports for which GLP compliance is claimed are audited by QA. This audit should be conducted at the final draft stage, when all raw data have been gathered and no more major changes are intended.

The aims of the audit of the final report should be to determine whether:

- the study was carried out in accordance with the study plan and SOPs;
- the study has been accurately and completely reported;
- the report contains all the elements required by GLP;
- the report is internally consistent; and
- the raw data are complete and in compliance with GLP.

QA may find it helpful to record the audit of the final report in a form that is sufficiently detailed to enable the audit to be reconstructed. Procedures must be established so that QA is made aware of all additions or changes made to the study data and report during the audit phase.

Before signing the QA statement, QA should ensure that all issues raised in the QA audit have been appropriately addressed in the final report, that all agreed actions have been completed, and that no changes to the report have been made which would require a further audit.

Any correction of or addition to a completed final report must be audited by QA. A revised or additional QA statement would then need to be provided.

The Quality Assurance statement

The Principles of GLP require that a signed quality assurance statement be included in the final report, which specifies types of inspections and their dates, including the phase(s) of study inspected, and the dates inspections results were reported to management and the Study Director and the Principal Investigator(s), if applicable [Sections II.2.2.1f) and II.9.2.4]. Procedures to ensure that this statement reflects QA's acceptance of the Study Director's GLP compliance statement and is relevant to the final study report as issued are the responsibility of management.

The format of the QA statement will be specific to the nature of the report. It is required that the statement include full study identification and the dates and phases of relevant QA monitoring activities. Where individual study-based inspections have not been part of the scheduled QA programme, a statement detailing the monitoring inspections that did take place must be included, for example, in the case of short-term studies where repeated inspections for each study are inefficient or impractical.

It is recommended that the QA statement only be completed if the Study Director's claim to GLP compliance can be supported. The QA statement would also serve to confirm that the final report reflects raw data. It remains the Study Director's responsibility to ensure that any areas of non-compliance with the GLP Principles are identified in the final report.

Quality Assurance and non-regulatory studies

Compliance with GLP is a regulatory requirement for the acceptance of certain studies. However, some test facilities conduct in the same area studies which are and which are not intended for submission to regulatory authorities. If the non-regulatory studies are not conducted in accordance with standards comparable to GLP, this will usually have a negative impact on the GLP compliance of regulatory studies.

Lists of studies kept by QA should identify both regulatory and non-regulatory studies to allow a proper assessment of workload, availability of facilities and possible interferences. QA should have access to an up-to-date copy of the master schedule to assist them in this task. It is not acceptable to claim GLP compliance for a non-GLP study after it has started. If a GLP-designated study is continued as a non-GLP study, this must be clearly documented.

Quality Assurance at small test facilities

At small test facilities, it may not be practicable for management to maintain personnel dedicated solely to QA. However, management must give at least one individual permanent, even if part-time, responsibility for co-ordination of the QA function. Some continuity in the QA staff is desirable to allow the accumulation of expertise and to ensure consistent interpretation. It is acceptable for individuals involved in studies that comply with GLP to perform the QA function for GLP studies conducted in other departments within the test facility. It is also acceptable for personnel from outside the test facility to undertake QA functions if the necessary effectiveness required to comply the GLP principles can be ensured.

This concept may be additionally applied to multi-site studies, for example field studies, on the condition that overall responsibility for co-ordination is clearly established.

2. COMPLIANCE OF LABORATORY SUPPLIERS WITH GOOD LABORATORY PRACTICE PRINCIPLES*

History

In the framework of the OECD Consensus Workshop on Good Laboratory Practice, held 16-18 October 1990 in Bad Dürkheim, Germany, a working group met to discuss and arrive at consensus on the compliance of laboratory suppliers with Principles of GLP. The Working Group was chaired by Dr. David Moore (Head, GLP Compliance Monitoring Authority, United Kingdom). Participants in the Working Group represented GLP compliance monitoring units and test facilities in Austria, Finland, France, Germany, Japan, Sweden and the United Kingdom.

The Working Group established the context of this consensus document, and made recommendations related to the role of suppliers vis-à-vis GLP Principles including the role of accreditation as a complementary tool to GLP compliance. It reached consensus and provided guidance on issues related to several specific categories of supplies. These issues are set out in the document.

The draft consensus document developed by the Working Group was circulated to member countries and revised, based on the comments received. It was subsequently endorsed by the OECD Panel on GLP and the Chemicals Group and Management Committee of the Special Programme on the Control of Chemicals. The Environment Committee then recommended that this document be derestricted under the authority of the Secretary-General.

In light of the adoption of the Revised OECD Principles of GLP in 1997, this Consensus Document was reviewed by the Working Group on GLP and revised to make it consistent with modifications made to the Principles. It was endorsed by the Working Group in April 1999 and, subsequently, by the Joint Meeting of the Chemicals Committee and Working Party on Chemicals, Pesticides and Biotechnology in August 1999. It was declassified under the authority of the Secretary-General.

Introduction

The responsibilities of the management of test facilities are defined in the OECD Principles of Good Laboratory Practice under the heading of Test Facility Organisation and Personnel (Section II.1). Test facility management should ensure that the GLP Principles are complied with at the test facility and that a sufficient number of qualified personnel, appropriate facilities, equipment and materials are available for the timely and proper conduct of the study. They also should ensure that test facility suppliers meet requirements appropriate to their use in a study. On the basis of these requirements, suppliers of materials used in studies submitted to regulatory authorities need not be

---

* No. 5 in the OECD Series on Principles of Good Laboratory Practice and Compliance Monitoring.

included in national GLP compliance programmes but they do play a definite role relating to the responsibilities of the management of test facilities.

As by definition in the GLP Principles, the responsibility for the quality and fitness for use of equipment and materials rests entirely with the management of the test facility. The acceptability of equipment and materials in GLP-compliant laboratories should therefore be guaranteed to any regulatory authority to whom studies are submitted. The main purpose of this document is to offer advice to both test facility management and suppliers as to how they might meet GLP requirements through national accreditation schemes and/ or working to formal national or international standards, or by adopting other measures which may be appropriate to a particular product. National or international standards, which may be set by an accreditation organisation, may be applied whenever they are acceptable to the test facility's management. The management of facilities, individually or in co-operation with each other, should thus maintain close contacts with suppliers and with their accreditation organisations.

Standards and accreditation schemes

Laboratories use various supplied materials in studies conducted in compliance with the GLP Principles. Suppliers have attempted to produce products which satisfy users' obligations as set out in the GLP Principles. Many suppliers have adopted manufacturing practices which comply with formal national or international standards, or have become accredited within various national schemes. These initiatives have been taken in the anticipation that supplied products will therefore be acceptable to regulatory authorities who require studies to be conducted in compliance with GLP Principles.

Suppliers are recommended to implement International Standard ISO 9001, and particularly Part 1 – Specification for Design/Development, Production, Installation and Servicing. This International Standard can be supported with European Standard EN 45001; the importance of Paragraph 5.4.7 of the latter, which refers to subcontracting, is emphasised.

Where appropriate, accreditation can be especially useful to suppliers. Accreditation schemes frequently monitor members' implementation of national and international standards; thus, a supplier or manufacturer's accreditation certificate may signify to the customer the satisfactory implementation of a standard in addition to other aspects of accreditation. It is recommended that suppliers seek membership, where feasible and/or appropriate, in national accreditation schemes.

Although accreditation is a useful complementary tool to support compliance with the GLP Principles, it is not an acceptable alternative to GLP compliance nor will it lead to international recognition in the context of meeting the requirements for the mutual acceptance of data as set out in the OECD Council Acts. (Decision of the Council concerning the Mutual Acceptance of Data in the Assessment of Chemicals [C(81)30(Final)], adopted 12 May 1981, and Council Decision-Recommendation on Compliance with the Principles of Good Laboratory Practice [C(89)87(Final)], adopted 2 October 1989. For the texts of both Council Acts, see the Annex, Sections 1 and 2.)

Test systems

The Revised Principles of GLP [Section II.8.2.5*b*)] require that the characterisation of test systems (animals, plants and other organisms) should be given in the study plan. This is the requirement that can be directly fulfilled by information from the supplier. In some countries where GLP has been implemented, suppliers belong to national regulatory or voluntary accreditation schemes (for example, for laboratory animals) which can provide users with additional documentary evidence that they are using a test system of a defined quality.

Animal feed, bedding and water

Although not specifically indicated in the Revised GLP Principles, animal feed should be analysed at regular intervals to establish its composition in order to avoid any potential interference with the test system. Water and bedding should also be analysed to ensure that contaminants are not present at levels capable of influencing the results of a study. Certificates of analysis are routinely provided by suppliers, including water authorities. Suppliers should provide appropriate documentary evidence to ensure the reliability of the analyses carried out.

Radio-labelled chemicals

Commercial pressure has forced suppliers of radio-labelled chemicals to seek formal GLP compliance by inclusion in national GLP compliance programmes. In many instances these suppliers produce labelled test items which are required to be fully characterised by procedures which comply with the GLP Principles. Suppliers of radio-labelled chemicals may need to be covered through national GLP compliance monitoring programmes.

Computer systems, applications software

All computer software, including that obtained from an external supplier, should normally be acceptance-tested before being put into service by a laboratory. From this requirement, it can be inferred that it is acceptable for formal validation of applications software to be carried out by the supplier on behalf of the user, provided that the user undertakes the formal acceptance tests.

The user should ensure that all software obtained externally has been provided by a recognised supplier. Many suppliers have endeavoured to meet users' requirements by implementing ISO 9001. This is considered to be useful.

The Revised Principles of GLP [Section II.1.2.2g)] place the responsibility to ensure that software programmes have been validated with the Study Director. The validation may be undertaken by the user or the supplier, but full documentation of the process must be available and should be retained in the archives. In cases where the validation is performed by the user, Standard Operating Procedures should be available [Section II.7.4.2b)].

It is the responsibility of the user to undertake an acceptance test before use of the software programme. The acceptance test should be fully documented.

[See OECD Consensus Document: The Application of the Principles of GLP to Computerised Systems, 1995, Part II, Section 6 of this publication.]

Reference items

It is the responsibility of test facility management to ensure that all manufactured reference items meet the GLP requirements for identity, composition, purity and stability for each batch of material (Sections II.6.2.2 and II.6.2.4 of the Revised Principles of GLP).

Certificates provided by suppliers should cover data on identity, purity and stability (under specified conditions if needed) and any other characteristics to define each batch appropriately. In special cases, the supplier may need to provide further information on, for example, methods of analysis, and should be prepared to demonstrate national/international measures of quality control, for example by reference to Good Manufacturing Practice or a national/international pharmacopoeia.

Apparatus

It is the responsibility of test facility management to ensure that instruments are adequate and functioning according to their intended use. Test facility management should also ensure that instruments are inspected and calibrated at prescribed intervals. Calibration should be traceable to national or international standards of measurement as appropriate. If reference standards are kept by the user, they should be calibrated by a competent body at prescribed intervals.

Suppliers are expected to provide all information necessary for the correct performance of the instruments. For certain types of instruments, for example balances and reference thermometers, calibration certificates should also be provided.

Sterilised materials

It is the responsibility of test facility management to ensure that materials which should be free from sources of infection have been properly sterilised with appropriate control procedures. Suppliers should be able to provide proper evidence, for example through certificates or reference to national standards, that materials sterilised by irradiation or other means or agents are free from sources of infection or undesirable residues from sterilisation agents.

General reagents

The user should ensure that reagents are obtained only from an accredited supplier. The supplier should provide documentary evidence of any accreditation status. Where there is no national accreditation scheme, the user should ensure receipt of a certificate of analysis from the supplier which guarantees that the reagent is as described by the label.

The user should be responsible for ensuring, by arrangement with the supplier, that all reagents are labelled with sufficient detail to comply with the specific requirements of GLP.

Detergents and disinfectants

The user should be aware of all active constituents to enable a suitable choice for use and to remove the potential for any contamination or interference which could be said to affect the integrity of a study.

Products required for microbiological testing

The user should be responsible for ensuring, by arrangement with the supplier, that all such products are labelled with at least the following information: source, identity, date of production, shelf life, storage conditions.

The supplier should ensure that documentation is available giving evidence of any accreditation status. Where there is no national accreditation scheme, the supplier should provide the user with a validation document which gives evidence of the fact that the product is as described by its label.

3. THE APPLICATION OF THE GOOD LABORATORY PRACTICE PRINCIPLES TO FIELD STUDIES*

History

In the framework of the Second OECD Consensus Workshop on Good Laboratory Practice, held 21-23 May 1991, in Vail, Colorado, experts discussed and reached consensus on the application of the GLP Principles to field studies. The Workshop was chaired by Dr. David Dull (Director, EPA Laboratory Data Integrity Program, United States). Experts from the following countries took part in the Consensus Workshop: Belgium, Canada, Denmark, Finland, Germany, the Netherlands, Switzerland, the United Kingdom and the United States.

The issues to be dealt with by the Workshop were defined at the First Consensus Workshop on GLP held in October 1990 in Bad Dürkheim, Germany. The Second Consensus Workshop was able to reach agreement on the management of field studies in relation to compliance with the GLP Principles, interpreting such concepts as study, test site, study director, management responsibilities, quality assurance, etc., for application in this specific context. The Consensus Document gives guidance for the interpretation of the relevant GLP Principles in relation to field studies.

The draft Consensus Document developed by the Second Consensus Workshop was circulated to member countries, and revised based on the comments received. It was subsequently endorsed by the OECD Panel on GLP and the Chemicals Group and Management Committee of the Special Programme on the Control of Chemicals. The Environment Committee then recommended that this document be derestricted under the authority of the Secretary-General.

In light of the adoption of the Revised OECD Principles of GLP in 1997, this Consensus Document was reviewed by the Working Group on GLP and revised to make it consistent with modifications made to the Principles. It was endorsed by the Working Group in June 1999 and, subsequently, by the Joint Meeting of the Chemicals Committee and Working Party on Chemicals, Pesticides and Biotechnology in August 1999. It is declassified under the authority of the Secretary-General.

Introduction

The Principles of Good Laboratory Practice (GLP), as adopted by the OECD in 1981 and revised in 1997, provide recommended test management standards for a wide variety of studies done for regulatory purposes or other assessment-related purposes. The report of the Expert Group ("Good Laboratory Practice in the Testing of Chemicals", OECD, 1982, out of print) which developed the GLP Principles in 1981, expressly lists the following types of tests as covered by the GLP Principles:

- physico-chemical properties;

---

* No. 6 in the OECD Series on Principles of Good Laboratory Practice and Compliance Monitoring.

- toxicological studies designed to evaluate human health effects (short- and long-term);
- ecotoxicological studies designed to evaluate environmental effects (short- and long-term); and
- ecological studies designed to evaluate environmental chemical fate (transport, biodegradation, and bioaccumulation).

Testing intended to determine the identity and magnitude of pesticide residues, metabolites, and related compounds for tolerance and other dietary exposure purposes is also included in the overall classification of ecological studies. The GLP Principles are intended to cover a broad range of commercial chemical products including pesticides, pharmaceuticals, cosmetics, veterinary drugs as well as food additives, feed additives and industrial chemicals.

Most experience in GLP compliance monitoring by the national monitoring authorities in OECD member countries has been gained in areas related to (non-clinical) toxicological testing. This is because these studies were traditionally deemed of greatest importance from a human health standpoint, and early identified laboratory problems primarily involved toxicological testing. Many established compliance monitoring procedures of the OECD member countries were thus developed from experience gained in the inspection of toxicology laboratories. Compliance monitoring procedures for laboratories performing ecotoxicological studies are also relatively well developed.

The area of field studies with pesticides or veterinary drugs, such as residue, metabolism, and ecological studies, presents a substantial challenge to GLP monitoring authorities and experimental testing facilities in that study plans, conditions, methods, techniques, and findings differ significantly from those traditionally associated with toxicological testing, as well as most laboratory-based ecotoxicological testing.

In the following, the special issues associated with field studies are identified and addressed in order to provide meaningful guidance and interpretation with respect to the Revised Principles of GLP. Many of the points in the original Consensus Document were integrated into the Revised Principles. The following deals only with those issues which might still be considered to need further interpretation.

Interpretations related to definitions of terms

The expression "non-clinical health and environmental safety study" in the definition of Good Laboratory Practice is understood to include field studies. A field study is a study which includes experimental activities carried out outside the usual laboratory situation, such as on land plots, in outdoor ponds or in greenhouses, often in combination or in sequence with activities carried out in a laboratory.

Field studies include, but are not limited to, studies for determining:

- magnitude of residue;
- photodegradation;
- plant metabolism;
- soil metabolism;
- rotational crop uptake;
- soil dissipation;
- effects on mesocosms;

- bioaccumulation; and
- effects on non-target organisms.

The term "test facility", when applied to field studies, may include several "test sites", at one or more geographical locations, where phases or components of a single overall study are conducted. The different test sites may include, but are not limited to:

- Research laboratory(ies) where test/reference item characterisation (including determination of identity, purity/strength, stability, and other related activities) is conducted.
- One or more agricultural or other in- or outdoor sites (like greenhouses) where the test or control item is applied to the test system.
- In some cases, a processing facility where harvested commodities are treated to prepare other items, *e.g.* the conversion of tomatoes into juice, puree, paste, or sauce.
- One or more laboratories where collected specimens (including specimens from processing) are analysed for chemical or biological residues, or are otherwise evaluated.

"Study Director" and "Principle Investigator": in field studies which could involve work at more than one test site, some of the Study Director's responsibilities may be delegated. At each test site when the Study Director cannot exercise immediate supervision, study procedures may be controlled by a member of the staff, called the Principal Investigator. The Principal Investigator means an individual responsible for the conduct of certain defined phases of the study, acting on behalf of the Study Director. The responsibilities of the Principal Investigator are described in the Revised GLP Principles in Section II.1 and in the section on "Principal Investigator's Responsibilities" below.

A "non-clinical health and environmental safety study" in the field, at one or more test sites, could include both the field and laboratory phases defined in a single "study plan".

"Test system" could also include complex ecological systems.

"Test item" could include, but need not be limited to: a chemical substance or mixture, a radio-labelled compound, a substance of biological origin, or a process waste. In the context of field residue or environmental studies, the test item is generally an *active ingredient* or a mixture (formulation) comprising active ingredient(s) and one or more inert components such as emulsifiers. Other field studies on plant and soil metabolism are designed to study the *fate* of the test item and use radio-labelled forms of the chemical; the test item can be *analytical grade* or *technical grade* material which may be formulated at the field site immediately prior to application.

In the context of field studies, "reference items" are also understood to include analytical standards. They should be adequately characterised for the type of study being conducted, and this characterisation should be addressed in the study plan.

In field studies, the term "vehicle" generally refers to the diluent, if any, used to dilute the test item (usually a formulation or a tank mix of a pesticide). The term also includes any additional solvents, surface active agents or other chemicals used to enhance the solubility or application characteristics.

Interpretations related to test facility organisation and personnel

Test facility management's responsibilities

Management, from the perspective of the GLP Principles, has several connotations and may involve several persons in several locations. The management level to which the Study Director reports has the *ultimate* responsibility for ensuring that the facilities operate in compliance with GLP Principles. In the context of field studies, there may also be several "test site management" entities that are primarily responsible for personnel, facilities, apparatus and materials at each test site and for formally assuring the Study Director (in writing) that these requirements can be met for the appropriate phase of each study. Test site management must also assure the Study Director that the provisions of the GLP Principles will be followed.

Test site management must assure the Study Director and his/her management that there is an appropriately qualified individual (Principal Investigator) at the test site who can effectively carry out his/her phase of the study in conformance with the study plan, applicable SOPs, the GLP Principles and the specific technical requirements. The overall management must have a firm understanding and working agreement with the test site management as to how and by whom the Quality Assurance Programme (QAP) will be carried out.

With multiple levels of management, study personnel and QAP staff, it is critical that there are clear lines of authority and communication, and assigned responsibilities, so that the Study Director can effectively carry out his/her GLP responsibilities. This should be documented in writing. It is the responsibility of the overall management to ensure that clear lines of communication exist.

There are likely to be some test sites where aspects of study conduct are indirectly (or directly) carried out by non-permanently employed personnel. Where these persons have generated or entered raw data, or have performed unsupervised activities relevant to the conduct of the study, records of their qualifications, training and experience should be maintained. Where these individuals have carried out routine maintenance operations such as crop thinning, weeding, fertilisation, etc. subject to supervision by more highly qualified staff, no such personnel records need be maintained.

Study Director's responsibilities

The designation of the Study Director is a key decision in assuring that a study will be properly conducted according to the GLP Principles. The terminology "responsibility for the overall conduct of the study and for its final report" may be interpreted in a broad sense for most field studies, as the Study Director may be geographically remote from parts of the actual experimental work. The Study Director thus will have to rely heavily on his/her designated Principal Investigator(s) and associated technical personnel at each test site to assure technical reliability and GLP compliance. The responsibilities of such personnel should be explicitly fixed in writing.

Effective communications have to be established and maintained between the Study Director and all associated personnel to ensure that the study plan and SOPs are being followed, and that all other GLP requirements are being met. Communications with participating QAP personnel are also critical to ensure that they are properly notified of critical phase activity, that QAP inspection reports are transmitted in a timely manner, and that corrective actions are implemented in a meaningful fashion.

As part of his/her duties, the Study Director has responsibility in ensuring that: 1) adequately characterised test and reference items are available at the test sites, as necessary; 2) there is adequate co-ordination between field (or processing) sites and analytical laboratories for specimen analyses; and 3) data from field, processing and laboratory sites are properly collated and archived.

Principal Investigator's responsibilities

Where a Study Director cannot exercise on-site supervisory control over any given phase of the study, a Principal Investigator will be identified/nominated to act on the Study Director's behalf for the defined phase.

The Principal Investigator will be named in the study plan or amendment, which will also delineate the phase(s) of the study covered by his responsibilities. The Principal Investigator will be an appropriately qualified and experienced individual suitably positioned to be able to immediately supervise the applicable phase.

The Principal Investigator, acting on behalf of the Study Director, will ensure that the relevant phase(s) of the study are conducted in accordance with the study plan, relevant SOPs, and with GLP. These responsibilities will include, but are not necessarily limited to:

a) Collaborate as appropriate with the Study Director and other study scientists in the drafting of the study plan.

b) Ensure that the study personnel are properly briefed, that such briefings are documented, and that copies of the study plan and relevant SOPs are freely accessible to personnel as necessary.

c) Ensure that all experimental data, including unanticipated responses of the test system, are accurately recorded.

d) Ensure that all deviations from SOPs and the study plan (unforeseen occurrences or inadvertent errors) are noted when they occur and that, where necessary, corrective action is immediately taken; these are recorded in the raw data. As soon as practicable, inform the Study Director of such deviations. Amendments to the study plan (permanent changes, modifications or revisions), however, must be approved in writing by the Study Director.

e) Ensure that all relevant raw data and records are adequately maintained to assure data integrity and that they are transferred in a timely way to the Study Director or as directed in the study plan.

f) Ensure that all samples and specimens taken during the relevant study phase(s) are adequately protected against confusion and deterioration during handling and storage. Ensure that these samples and specimens are dispatched in an appropriate manner.

g) Sign and date a report of the relevant phase(s), certifying that the report accurately presents all the work done, and all the results obtained, and that the work was conducted in compliance with GLP. Include in this report sufficient commentary to enable the Study Director to write a valid Final Report covering the whole study, and send the report to the Study Director. The Principal Investigator may present the original raw data as his report, where applicable, including a statement of compliance with GLP.

Interpretations related to the Quality Assurance Programme

Usually, a single individual will not be able to perform the quality assurance function for field studies, but rather there will be a need for a number of persons. In some cases, these persons may all be in the employment of a single unit (for example, that of the study sponsor); in other cases, they may be employed by different units (for example, part by the study sponsor and part by contractors). There must be a full, frank flow of information from the different quality assurance persons to the responsible test site management, to the responsible Principal Investigator(s), to the Study Director as the person responsible for the overall conduct of the study, to the Study Director's management, and to the latter's Quality Assurance Programme. Likewise, it will be necessary to assure effective communications from the Study Director and/or Principal Investigators to the quality assurance personnel for notification of critical activities.

Because of the complex nature of field studies, which may involve similar activities at separate locations, and the fact that the exact time of certain activities will depend upon local weather or other conditions, flexible quality assurance procedures may be required. [See "Quality Assurance and GLP", Part II, page 25, in this publication.]

The geographical spread of test sites may mean that quality assurance personnel will also need to manage language differences in order to communicate with local study personnel, the Study Director, Principal Investigators and test site management.

Irrespective of where the test sites are located, the written reports of quality assurance personnel must reach both management and the Study Director. The actual receipt of such reports by management and the Study Director should be documented in the raw data.

Interpretations related to facilities

General

Facilities for a field study will typically consist wholly or partially of agricultural or farming units, forested areas, mesocosms or other outdoor study areas where there is customarily much less, or even no, control over the environmental conditions than that achievable in an enclosed laboratory or a greenhouse. Also, security and oversight of operations and facilities are not as manageable as for a laboratory-based study.

An issue of concern in pesticide field studies is the potential for contamination of the study plots from drift or overspray of pesticides being used on neighbouring property. This can particularly be a problem for test plots located in the midst of, or adjacent to, other land used for commercial agricultural activities. Study plot locations should be chosen so as to ensure minimal possibility of off-site interferences. Preferably, the plots should be located in areas free of interfering chemicals or where the historical pesticide use (both study and normal use applications) has been documented.

It is recognised that laboratories conducting pesticide residue analysis must be especially cognisant of the potential for contaminating specimens, as well as of reference standards. Receipt and storage areas for specimens must be separate from storage areas for pesticide formulations and other test or reference items. Areas used for specimen and sample preparation, instrumentation, calibration of sprays, reference standard preparation, and for washing glassware should be adequately isolated from each other and from other functions of the laboratory which might introduce contamination.

Facilities for handling test and reference items

Storage areas for test and reference items at all test sites should be environmentally monitored, if required, to assure conformance with established stability limits for these materials. Test and reference items should not be placed in the same storage containers with collected test system specimens and other materials of low concentrations which are being stored for shipment to the analytical laboratory or to off-site archives. There should be adequate storage and disposal facilities available for pesticide and related wastes such that there is no potential for cross-contamination of test systems, of test or reference items or of collected specimens.

Waste disposal

Of particular concern at field sites is the storage and disposal of excess pesticide dilutions (or tank mixes). The minimum volume of such dilutions should be prepared. In addition to assuring that these potentially hazardous wastes are not endangering human health or the environment, these materials also need to be controlled in such a way that there is no impact on test systems, specimens or other materials or equipment used in studies. It should also be assured that unused test and reference items are returned to the sponsors or suppliers, or are disposed of in a legal and responsible manner.

Interpretations related to apparatus, material and reagents

In the field phase, the frequency of operations such as inspection, cleaning, maintenance and calibration may need to reflect possible transport of the equipment (for example when balances are moved from site to site). These operations should be described by Standard Operating Procedures.

Apparatus which is used only for one specific study (*e.g.* leased or rented equipment, or equipment such as sprayers which have been specifically configured for use in one study) may not have records of periodic inspection, cleaning, maintenance and calibration. In such cases, this information may be recorded in the study-specific raw data. If it is not feasible to document the relevant procedures as SOPs, they can be documented in study plans, with references to handbooks.

Materials and reagents should be verified as being non-interfering by the analysis of an adequate number of "reagent blanks".

Interpretations related to test systems

Some test systems utilised in field studies may consist of complex ecosystems that will be difficult to characterise, identify or otherwise document to the extent that can be accomplished for more traditional test systems. However, these more complex test systems should be described by location and characteristics, to the degree possible, in the study plan, and the actual study plot areas identified by signs, markers or other means. Plants, seeds, soils and other materials being used as test systems should be described and documented as to their source, date(s) of acquisition, variety, strain, cultivar or other identifying characteristics, as appropriate. Soil should be characterised to the degree necessary and documented to verify suitability for its use in field studies.

As noted under "Facilities", test systems for pesticide studies should be free from interferences from outside sources, particularly drift or overspray from neighbouring plots. If relevant, the study plan should discuss the need for analysis of preliminary or pre-treatment control samples. Control plots and buffer zones are to be used to the degree necessary to account for or minimise potential interferences or other forms of study bias.

Interpretations related to test and reference items

Receipt, handling, sampling and storage

The following documentation should be present at the test site:

- Source, *e.g.* commercial formulation, special formulation, etc.
- Mode of transfer, with retention of shipping documents.
- Date of receipt.
- Condition of substance on receipt.
- Storage location and conditions.
- Complete log documenting distribution, accounting for the total amount of the test item and final disposal.

Characterisation

It is not necessary to have all characterisation records and data available at each test site. However, sufficient information needs to be present to assure that the test and reference items have been adequately characterised. This generally will comprise: name of the chemical (*e.g.* CAS number, code name, etc.); lot or batch number; amount of active ingredient; site where the analyses were conducted, and where the relevant raw data are archived; stability with regard to storage and transfer conditions (*i.e.* expiry date, temperature range); and safety precautions.

Product chemistry data based on separate laboratory experiments will frequently have defined the stability of test item mixtures in the vehicle over a range of pH, temperature and hardness values. If relevant restrictions are known, then the study plan may specify appropriate ranges for the application, and the actual values should be recorded in the raw data as well as the time of mixing and the termination of the application.

Similar data for homogeneity are also often available from producers that show non-separation of mixture phases over various periods of time under specified conditions.

If tank mix samples are to be analysed, this requirement should be specified in the study plan, along with sampling and analytical methodology.

Interpretations related to Standard Operating Procedures

Special emphasis should be placed on key procedures for field studies, such as test item storage, data collection in the field, application equipment calibration, test item application, and specimen collection and transportation.

The study plan will also require inclusion of all methodologies intended to be used for specimen analyses. This may require an approved study plan amendment if the method has not been fully developed or validated at the time the original study plan is signed. The study plan should also provide for all speciality analysis, *e.g.* confirmation procedures.

Interpretations related to performance of the study

Study plan

Study plans intended for most field studies will need to reflect more flexibility than traditional laboratory studies due to the unpredictable nature of the weather, the possibility of the need to employ borrowed or rented equipment, special arrangements for the preservation, storage and transport of specimen samples, or other special circumstances. Rather than citing specific dates in the study plan for key phases such as test item application, culturing operations and specimen sampling, a more realistic approach would be to specify commodity growth stages for these activities to the degree possible and giving only approximate time frames.

In order to approve study plan amendments in a timely and effective fashion, special communication procedures will need to be established between the personnel at the test sites and the Study Director if the two entities are not at the same location.

Conduct of the Study

In view of the importance of quality control measures in residue and environmental analyses, these should be addressed in SOPs and/or in the study plan. Procedures to evaluate reproducibility, freedom from interferences, and confirmation of analytic identity would typically be included.

Raw data includes any worksheets, records, memoranda, notes, or exact copies thereof that are the result of original observations and activities of a study and are necessary for the reconstruction and evaluation of the report of that study. In the event that exact transcripts of raw data have been prepared (*e.g.* tapes which have been transcribed verbatim, dated, and verified accurate by signature), the exact copy or exact transcript may be substituted for the original source as raw data. Examples of raw data include photographs, microfilm, or microfiche copies, computer printouts, magnetic media, including dictated observations, and recorded data from automated instruments.

It is recommended that all entries be made with indelible ink. Under some circumstances, use of pencil in the field may be unavoidable. When this is necessary, "verified" copies should be prepared as soon as practicable. Any entries in pencil or in different colours should be appropriately identified on the verified copies. In addition, study records should clearly state the reason for using pencil.

Interpretations related to reporting of study results

The report(s) of the Principal Investigator(s) can be attached to the overall study report by the Study Director as appendices as described in Paragraph *g)* in the note under Principal Investigator's responsibilities, above.

Interpretations related to storage and retention of records and materials

One potential problem area associated with remote test sites is the temporary storage of materials from ongoing studies until they can be transferred to archives at the end of the study. Temporary storage facilities at all test sites should be adequate to ensure the integrity of the study materials.

4. THE APPLICATION OF THE GOOD LABORATORY PRACTICE PRINCIPLES TO SHORT-TERM STUDIES*

History

In the framework of the Third OECD Consensus Workshop on Good Laboratory Practice held 5-8 October 1992 in Interlaken, Switzerland, a working group of experts discussed the interpretation of the GLP Principles as applied to short-term studies. This working group was chaired by Ms. Francisca E. Liem (United States Environmental Protection Agency); the rapporteur was Dr. Hans-Wilhelm Hembeck (German GLP Federal Office). Participants in the working group were from both national GLP compliance monitoring authorities and from testing laboratories in the following countries: Australia, Austria, the Czech Republic, Finland, France, Germany, Ireland, the Netherlands, Poland, Sweden, Switzerland, the United Kingdom and the United States. Two sub-working groups were formed and chaired by Ms. Liem (short-term biological studies) and Dr. Hembeck (physical-chemical studies); the respective rapporteurs were Mr. David Long (France) and Dr. Stephen Harston (Germany). The document developed by the working group cites the appropriate OECD Principles of GLP and gives guidance on their interpretation in relation to short-term studies in a series of notes.

The draft document developed by the working group was circulated to member countries for comments. The text was revised, based on comments received, and reviewed by the OECD Panel on Good Laboratory Practice at its fifth meeting in March 1993, which amended the text and forwarded it to the Joint Meeting of the Chemicals Group and Management Committee of the Special Programme on the Control of Chemicals. At its 20th Session, the Joint Meeting endorsed the document with minor editorial changes and recommended that it be derestricted under the authority of the Secretary-General.

In light of the adoption of the Revised OECD Principles of GLP in 1997, this Consensus Document was reviewed by the Working Group on GLP and revised to make it consistent with modifications made to the Principles. It was endorsed by the Working Group in April 1999 and subsequently by the Joint Meeting of the Chemicals Committee and Working Party on Chemicals, Pesticides and Biotechnology in August 1999. It is declassified under the authority of the Secretary-General.

Introduction

The OECD Principles of GLP are general and not specific to any particular type of test or testing discipline. The initial experience in OECD member countries in compliance monitoring has been primarily in long-term toxicity studies. Although subject to the OECD Principles of GLP, short-term studies present special concerns to management and compliance monitoring authorities based upon the existence of particular procedures and techniques.

---

* No. 7 in the OECD Series on Principles of Good Laboratory Practice and Compliance Monitoring.

The Revised Principles of GLP define a short-term study as "a study of short duration with widely used, routine techniques" (I.2.3.2). Short-term biological studies include acute toxicity studies, some mutagenicity studies, and acute ecotoxicological studies.

Physical-chemical studies are those studies, tests or measurements which are of a short duration (typically not more than one working week), employ widely-used techniques (*e.g.* OECD Test Guidelines) and yield easily repeatable results, often expressed by simple numerical values or verbal expressions.

Typical physical-chemical studies include, but are not limited to, chemical characterisation studies, melting point, vapour pressure, partition coefficient, explosive properties and other similar studies for which test guidelines exist. However, the regulatory agencies/receiving authorities in member countries will specify which of these tests should be submitted to them and which should be conducted under the Principles of GLP.

NOTES TO THE GLP PRINCIPLES

The following paragraphs of the Revised OECD Principles of Good Laboratory Practice need interpretation for their application to short-term studies. Paragraphs of the Revised OECD Principles which do not require interpretation are not repeated here. Notes are given for further guidance and interpretation.

II.1. TEST FACILITY ORGANISATION AND PERSONNEL

II.1.2. *Test facility management's responsibilities*

II.1.2.g) (Test facility management should) ensure that, for each study, an individual with the appropriate qualifications, training, and experience is designated by the management as the Study Director before the study is initiated…

[NOTE]: The designation of the Study Director is a key decision in assuring that the study will be properly planned, conducted and reported. The appropriate Study Director qualifications may be based more on experience than on advanced education.

II.2. QUALITY ASSURANCE PROGRAMME

II.2.1. *General*

II.2.1.1. The test facility should have a documented Quality Assurance Programme to assure that studies performed are in compliance with these Principles of Good Laboratory Practice.

[NOTE 1]: All references to "quality assurance programme" in this document should be interpreted with reference to the OECD Principles of GLP and the "OECD Consensus Document on Quality Assurance and Good Laboratory Practice" (Part II, Chapter 1 of this publication). In respect of physical-chemical studies, it is recognised that other published standards (*e.g.* ISO 9000 series) use the term "quality assurance" in a different way.

[NOTE 2]: The documentation of the Quality Assurance Programme should include a description of the use made of "study-based", "facility-based" or "process-based" inspections as defined in the OECD Consensus Document No. 4 "Quality Assurance and GLP". These definitions are reproduced below:

"*Study-based inspections*: These are scheduled according to the chronology of a given study, usually by first identifying the critical phases of the study.

Facility-based inspections: These are not based upon specific studies, but cover the general facilities and activities within a laboratory (installations, support services, computer system, training, environmental monitoring, maintenance, calibration, etc.).

Process-based inspections: Again, these are performed independently of specific studies. They are conducted to monitor procedures or processes of a repetitive nature and are generally performed on a random basis. These inspections take place when a process is undertaken very frequently within a laboratory and it is therefore considered inefficient or impractical to undertake study-based inspections. It is recognised that performance of process-based inspections covering phases which occur with a very high frequency may result in some studies not being inspected on an individual basis during their experimental phases."

II.2.2. Responsibilities of the Quality Assurance personnel

II.2.2.1. The responsibilities of the Quality Assurance personnel include, but are not limited to, the following functions. They should:

   a) Maintain copies of all approved study plans and Standard Operating Procedures in use in the test facility and have access to an up-to-date copy of the master schedule.

   b) Verify that the study plan contains the information required for compliance with these Principles of Good Laboratory Practice. This verification should be documented.

   c) Conduct inspections to determine if all studies are conducted in accordance with these Principles of Good Laboratory Practice. Inspections should also determine that study plans and Standard Operating Procedures have been made available to study personnel and are being followed.

[NOTE]: Because of the high frequency and routine nature of some standard short-term studies, it is recognised in the "OECD Consensus Document on Quality Assurance and Good Laboratory Practice" that each study need not be inspected individually by Quality Assurance during the experimental phase of the study. In these circumstances, a process-based inspection programme may cover each study type. The frequency of such inspections should be specified in approved Quality Assurance Standard Operating Procedures, taking into account the numbers, frequency and/or complexity of the studies being conducted in the facility. The frequency of inspections should be specified in the relevant QA Standard Operating Procedures, and there should be SOPs to ensure that all such processes are inspected on regular basis.

   f) Prepare and sign a statement, to be included with the final report, which specifies types of inspections and their dates, including the phase(s) of the study inspected, and the dates inspection results were reported to management and the Study Director and Principal Investigator(s), if applicable. This statement would also serve to confirm that the final report reflects the raw data.

[NOTE]: Where individual study-based inspections did not take place, the QA-statement must clearly describe which types of inspections (*e.g.* process-based) were performed and when. The QA-statement must indicate that the final report was audited.

II.3. FACILITIES

II.3.1. General

II.3.1.1. The test facility should be of suitable size, construction and location to meet the requirements of the study and to minimise disturbances that would interfere with the validity of the study.

II.3.1.2. The design of the test facility should provide an adequate degree of separation of the different activities to assure the proper conduct of each study.

[NOTE]: The issue of concern, primarily for biological in vitro studies is the possibility of contamination of the test system. Laboratories should establish facilities and procedures which demonstrably prevent and/or control such potential contamination.

II.4. APPARATUS, MATERIAL AND REAGENTS

II.4.2. Apparatus used in a study should be periodically inspected, cleaned, maintained, and calibrated according to Standard Operating Procedures. Records of these activities should be maintained. Calibration should, where appropriate, be traceable to national or international standards of measurement.

[NOTE]: Calibration should, where appropriate, provide for traceability of measurements to fundamental physical quantities maintained by appropriate national authorities. Apparatus should be checked periodically for continuing accuracy of measurement. Calibration substances should be treated as reference items, but need not be retained.

II.5. TEST SYSTEMS

II.5.1. Physical/chemical

[NOTE]: There is overlap between the requirements for "Physical/chemical test systems" in Section II.5.1.1 of the Revised OECD GLP Principles and those for apparatus in Section II.4.1 This overlap seems to have no practical implications for studies of this type. Apparatus used in a physical/chemical test system should be periodically inspected, cleaned, maintained, and calibrated according to SOPs, as specified above (Section II.4 of the Revised GLP Principles).

II.5.2. Biological

II.5.2.1. Proper conditions should be established and maintained for the storage, housing, handling and care of biological test systems, in order to ensure the quality of the data.

II.5.2.2. Newly received animal and plant test systems should be isolated until their health status has been evaluated. If any unusual mortality or morbidity occurs, this lot should not be used in studies and, when appropriate, should be humanely destroyed. At the experimental starting date of a study, test systems should be free of any disease or condition that might interfere with the purpose or conduct of the study. Test systems that become diseased or injured during the course of a study should be isolated and treated, if necessary to maintain the integrity of the study. Any diagnosis and treatment of any disease before or during a study should be recorded.

II.5.2.3. Records of source, date of arrival, and arrival condition of test systems should be maintained.

II.5.2.4. Biological test systems should be acclimatised to the test environment for an adequate period before the first administration/application of the test or reference item.

II.5.2.5. All information needed to properly identify the test systems should appear on their housing or containers. Individual test systems that are to be removed from their housing or containers during the conduct of the study should bear appropriate identification, wherever possible.

II.5.2.6. During use, housing or containers for test systems should be cleaned and sanitised at appropriate intervals. Any material that comes into contact with the test system should be free of contaminants at levels that would interfere with the study. Bedding for animals should be changed as required by sound husbandry practice. Use of pest control agents should be documented.

[NOTE 1]: Test system information: Record keeping is required to document the growth, vitality and absence of contamination of batches of in vitro test systems. It is important that the origin, substrain and maintenance of the test system be identified and recorded for in vitro studies.

[NOTE 2]: Characterisation of the test system, primarily for in vitro studies: It is essential that there is assurance that the test system as described in the study plan is being used, and is free of contamination. This can be accomplished, for example, by periodically testing for genetic markers, karyotypes, or testing for mycoplasma.

[NOTE 3]: Isolation of test systems: In the case of short-term biological studies, isolation of animal and plant test systems may not be required. The test facility SOPs should define the system for health status evaluation (e.g. historical colony and supplier information, observations, serological evaluation) and subsequent actions.

[NOTE 4]: Control of interfering materials in in vitro studies: There should be assurance that water, glassware and other laboratory equipment are free of substances which could interfere with the conduct of the test. Control groups should be included in the study plan to meet this objective. Periodic systems tests may also be performed to complement this goal.

[NOTE 5]: Characterisation of culture media: The types of media, ingredients and lot numbers of the media (e.g. antibiotics, serum, etc.) should be documented. Standard Operating Procedures should address the preparation and acceptance of such media.

[NOTE 6]: Test system use: Under certain circumstances, some member countries will accept the re-use of an animal or the simultaneous testing of multiple test items on one animal. The GLP issue of concern is that in all cases, complete historical documentation on the former use of the animal must be maintained and be referenced in the final report. It must also be documented that these practices do not interfere with the evaluation of the test item(s).

II.6. TEST AND REFERENCE ITEMS

II.6.2. Characterisation

II.6.2.1. Each test and reference item should be appropriately identified (*e.g.* code, Chemical Abstracts Service Registry Number [CAS number], name, biological parameters).

II.6.2.2. For each study, the identity, including batch number, purity, composition, concentrations, or other characteristics to appropriately define each batch of the test or reference items should be known.

II.6.2.3. In cases where the test item is supplied by the sponsor, there should be a mechanism, developed in co-operation between the sponsor and the test facility, to verify the identity of the test item subject to the study.

II.6.2.4. The stability of test and reference items under storage and test conditions should be known for all studies.

II.6.2.5. If the test item is administered or applied in a vehicle, the homogeneity, concentration and stability of the test item in that vehicle should be determined. For test items used in field studies (*e.g.* tank mixes), these may be determined through separate laboratory experiments.

II.6.2.6. A sample for analytical purposes from each batch of test item should be retained for all studies, except short-term studies.

[NOTE 1]: Adequate characterisation information should be available for each batch of the test and reference items. To promote acceptability in all member countries, it is recommended that this information should be generated in compliance with the Revised Principles of GLP when needed. Where the test item is in an early stage of development, it is acceptable for the analytical characterisation to be performed after the conduct of the biological study. However, there should be some information on the chemical structure of the test item before the study initiation date.

[NOTE 2]: To promote acceptability in all member countries, it is recommended that the stability of the test and reference items under conditions of storage should be determined in compliance with Principles of GLP when needed.

[NOTE 3]: There are considerable differences between the requirements of member countries concerning the evaluation of the concentration, stability and homogeneity of the test item in a vehicle. In addition, for certain short-term biological tests, it is not always possible to conduct such analyses concomitantly. For certain of these tests, if the time interval between preparation and application of a usually stable substance is only a few minutes, it might not be relevant to determine the stability of the test item. For these reasons, it is essential that analytical requirements are specified and approved in the study plan and clearly addressed in the final report.

[NOTE 4]: The data related to points II.6.2.4 and II.6.2.5 under "Characterisation" of test and reference items in the GLP Principles (above) may not be known in the case of physical-chemical studies being conducted to determine such data.

II.7 STANDARD OPERATING PROCEDURES

[NOTE]: The illustrative examples given in the Section II.7.4.4 of the Revised Principles of GLP (Test system) refer mainly to biological test systems and may thus not be relevant in the context of physical-chemical studies. It is the responsibility of test facility management to ensure that appropriate Standard Operating Procedures are produced for the studies performed in the facilities.

II.8. PERFORMANCE OF THE STUDY

II.8.1. Study plan

II.8.1.1. For each study, a written plan should exist prior to the initiation of the study. The study plan should be approved by dated signature of the Study Director and verified for GLP compliance by Quality Assurance personnel as specified in Section II.2.2.1b) above. The study plan should also be approved by the test facility management and the sponsor, if required by national regulation or legislation in the country where the study is being performed.

II.8.1.3. For short-term studies, a general study plan accompanied by a study specific supplement may be used.

[NOTE]: Where a particular short-term study or a series of such studies is performed frequently within a laboratory, it may be appropriate to prepare a single general study plan containing the majority of general information required in such a plan and approved in advance by the testing facility management and by the Study Director(s) responsible for the conduct of such studies and by QA.

Study-specific supplements to such plans (e.g. with details on test item, experimental starting date) should then be issued as a supplementary document requiring only the dated signature of the designated Study Director. The combined document – the general study plan and the study-specific supplement – is the study plan. It is important that such supplements are provided promptly to test facility management and to QA assurance personnel.

II.8.2. Content of the study plan

[NOTE]: The contents of the complete study plan (that is, of the general study plan and the study-specific supplement) should be as described in the Revised OECD Principles of GLP, with the possible exceptions noted below.

The study plan should contain, but not be limited to, the following information:

II.8.2.1. *Identification of the study, the test item and reference item*

   a) A descriptive title.

   b) A statement which reveals the nature and purpose of the study.

[NOTE]: This may not be needed if this information is provided by the descriptive title.

   c) Identification of the test item by code or name (IUPAC; CAS number, biological parameters, etc).

   d) The reference item to be used.

II.8.2.5. *Issues (where applicable)*

   a) The justification for selection of the test system.

   b) Characterisation of the test system, such as the species, strain, substrain, source of supply, number, body weight range, sex, age, and other pertinent information.

   c) The method of administration and the reason for its choice.

   d) The dose levels and/or concentration(s), frequency, and duration of administration/application.

[NOTE]: Issues a)-d), above, may not be needed for physical-chemical studies.

e) Detailed information on the experimental design, including a description of the chronological procedure of the study, all methods, materials and conditions, type and frequency of analysis, measurements, observations and examinations to be performed, and statistical methods to be used (if any).

[NOTE]: This may generally be given in a brief, summary form, or with reference to appropriate SOPs or Test Guidelines.

II.9. REPORTING OF STUDY RESULTS

II.9.1. General

II.9.1.1. A final report should be prepared for each study. In the case of short-term studies, a standardised final report accompanied by a study-specific extension may be prepared.

[NOTE]: Where short-term studies are performed using general study plans, it may also be appropriate to issue standardised final reports containing the majority of general information required in such reports and authorised in advance by the testing facility management, and by the Study Director(s) responsible for the conduct of such studies. Study-specific extensions to such reports (e.g. with details of the test item and the numerical results obtained) may then be issued as a supplementary document requiring only the dated signature of the Study Director. It is not acceptable to utilise a standardised final report when the study plan is revised or amended prior to or during the conduct of the study unless the standardised final report is amended correspondingly.

II.9.2. Content of the final report

[NOTE]: The contents of the complete final report (that is, of the "standardised final report" and the study-specific supplement) should be as described in the Revised OECD Principles of GLP, with the possible exceptions noted below:

The final report should include, but not be limited to, the following information:

II.9.2.1. *Identification of the study, the test and reference item*

   a) A descriptive title.
   b) Identification of the test item by code or name (IUPAC; CAS number, biological parameters, etc.).
   c) Identification of the reference item by chemical name.
   d) Characterisation of the test item including purity, stability and homogeneity.

[NOTE]: This may not be relevant when the study is carried out to determine such data.

II.9.2.4. *Statement*

A Quality Assurance Programme statement listing the types of inspections made and their dates, including the phase(s) inspected, and the dates any inspection results were reported to management and to the Study Director and Principal Investigator(s), if applicable. This statement would also serve to confirm that the final report reflects the raw data.

[NOTE]: This may need to reflect the use of process-based inspection. The QA Statement must clearly indicate that the final report was audited. [See also the note under "Responsibilities of the Quality Assurance personnel", II.2.2.1f), above.]

5. THE ROLE AND RESPONSIBILITIES OF THE STUDY DIRECTOR IN GOOD LABORATORY PRACTICE STUDIES*

History

In the framework of the Third OECD Consensus Workshop on Good Laboratory Practice held 5-8 October 1992 in Interlaken, Switzerland, a working group of experts discussed the interpretation of the GLP Principles as applied to the role and responsibilities of the Study Director. This working group was chaired by Dr. David F. Moore of the United Kingdom GLP Compliance Monitoring Authority; the Rapporteur was Dr. Heinz Reust (Swiss Federal Office of Public Health). Participants in the Working Group were from both national GLP compliance monitoring authorities and from testing laboratories in the following countries: Austria, Canada, the Federation of Russia, Finland, Germany, Japan, the Netherlands, Switzerland, the United Kingdom and the United States.

The draft document developed by the working group was circulated to member countries for comments. The text was revised, based on comments received, and reviewed by the OECD Panel on Good Laboratory Practice at its fifth meeting in March 1993, which amended the text and forwarded it to the Joint Meeting of the Chemicals Group and Management Committee of the Special Programme on the Control of Chemicals. At its 20th Session, the Joint Meeting endorsed the document with minor editorial changes and recommended that it be derestricted under the authority of the Secretary-General.

In light of the adoption of the Revised OECD Principles of GLP in 1997, this Consensus Document was reviewed by the Working Group on GLP and revised to make it consistent with modifications made to the Principles. It was endorsed by the Working Group in April 1999 and, subsequently, by the Joint Meeting of the Chemicals Committee and Working Party on Chemicals, Pesticides and Biotechnology in August 1999. It is declassified under the authority of the Secretary-General.

The role of the Study Director

The Study Director represents the single point of study control with ultimate responsibility for the overall scientific conduct of the study. This is the prime role of the Study Director, and all duties and responsibilities as outlined in the GLP Principles stem from it. Experience has shown that unless responsibility for the proper conduct of a study is assigned to *one* person, there is a potential for personnel to receive conflicting instructions, which can result in poor implementation of the study plan. There can be only one Study Director for a study at any given time. Although some of the *duties* of the Study Director can be delegated, as in the case of a subcontracted study, the ultimate responsibility of the Study Director as the single central point of control cannot.

---

* No. 8 in the OECD Series on Principles of Good Laboratory Practice and Compliance Monitoring.

In this regard, the Study Director serves to assure that the scientific, administrative and regulatory aspects of the study are controlled. The Study Director accomplishes this by coordinating the inputs of management, scientific/technical staff and the Quality Assurance Programme.

In multi-site studies which involve work at more than one test site and the Study Director cannot exercise immediate supervision, study procedures may be controlled by an appropriately trained, qualified and experienced member of the staff, called the Principal Investigator. He is responsible for the conduct of certain defined phases of the study in accordance with the applicable Principles of Good Laboratory Practice, acting on behalf of the Study Director.

Scientifically, the Study Director is usually the scientist responsible for study plan design and approval, as well as overseeing data collection, analysis and reporting. The Study Director is responsible for drawing the final overall conclusions from the study. As the lead scientist, the Study Director must coordinate with other study scientists, and/or Principal Investigator(s) keeping informed of their findings during the study and receiving and evaluating their respective individual reports for inclusion in the final study report.

Administratively, the Study Director must request and coordinate resources provided by management, such as personnel, equipment and facilities, to ensure they are adequate and available as scheduled for the proper conduct of the study.

Compliance with regulations is also the responsibility of the Study Director. In this role, the Study Director is responsible for ensuring that the study is carried out in accordance with the Principles of GLP, which require the Study Director's signature on the final study report to confirm compliance with the GLP Principles.

Management responsibilities

Management of a testing facility is responsible for ensuring that the facility operates in compliance with GLP Principles. This responsibility includes the appointment and effective organisation of an adequate number of appropriately qualified and experienced staff throughout the facility, including Study Directors, and, in the event of multi-site studies, Principal Investigator(s), if needed.

Appointment of Study Directors

Management should maintain a policy document defining the procedures adopted for selection and appointment of Study Directors, their deputies, and Principal Investigator(s) if required by national programmes.

When appointing a Study Director to a study, management should be aware of that person's current or anticipated workloads. The master schedule, which includes information on the type and timing of studies allocated to each Study Director, can be used to assess the volume of work being performed by individuals within the testing facility and is a useful management tool when allocating studies.

Replacement of a Study Director and/or Principal Investigator should be done according to established procedures and should be documented.

Training of Study Directors

Management should ensure that there is documentation of training in all aspects of the Study Director's work. A training programme should ensure that Study Directors have a thorough understanding of GLP Principles and an appropriate knowledge of testing facility procedures. This may include an awareness and working knowledge of other guidelines and regulations pertinent to the testing facility and the particular study type, for example, the OECD Test Guidelines. Training may include work experience under the supervision of competent staff. Observation periods or work experience within each discipline involved in a study can provide a useful basic understanding of relevant practical aspects and scientific principles, and assist in the formation of communication links. Attendance at in-house and external seminars and courses, membership in professional societies and access to appropriate literature may allow Study Directors to maintain current awareness of developments within their scientific field. Professional development should be continuous and subject to periodic review. All training should be documented and records should be retained for the period specified by the appropriate authorities.

Documented records of such a programme should reflect the progression of training and provide a clear indication of the type of study that an individual is considered competent to direct. Further training or retraining may be necessary from time to time, for example, following the introduction of new technology, procedures or regulatory requirements.

Responsibilities of the Study Director

The Study Director is the individual who has overall responsibility for the scientific conduct of a study and can confirm the compliance of the study with the OECD Principles of Good Laboratory Practice.

Study initiation

The Study Director has to approve the study plan, which is prepared before study initiation by dated signature. This document should clearly define the objectives and the whole conduct of the study and how they are to be achieved. Any amendments to the study plan have to be approved as mentioned above. For a multi-site study, the study plan should identify and define the role of any Principal Investigator(s) and any test facilities and test sites involved in the conduct of the study.

The Study Director should take responsibility for the study by dated signature of the study plan, at which stage the study plan becomes the official working document for that study (study initiation date). If appropriate, the Study Director should also ensure that the study plan has been signed by the sponsor and the management, if required by national programmes.

Before the study initiation date, the Study Director should make the study plan available to Quality Assurance (QA) staff for verifying that it contains all information required for compliance with the GLP Principles.

Before the experimental starting date of the study, the Study Director should assure that copies of the study plan are supplied to all personnel involved in the study; this should include Quality Assurance (QA) staff.

Before any work on the study is undertaken, the Study Director should ascertain that management have committed adequate resources to perform the study, and that adequate test materials and test systems are available.

Study conduct

The Study Director has responsibility for the overall conduct of the study and should ensure that the procedures laid down in the study plan including amendments are followed and all data generated during the study are fully documented. Specific technical responsibilities may be delegated to competent staff, and need to be documented.

The Study Director's involvement during the course of the study should include overviewing the study procedures and data to ensure that the procedures laid down in the study plan are being followed and that there is compliance with the relevant Standard Operating Procedures, and should include computer-generated data. In order to demonstrate this, the type and frequency of the reviews should be documented in the study records.

As all decisions that may affect the integrity of the study should ultimately be approved by the Study Director, it is important that he remains aware of the progress of the study. This is of particular importance following temporary absence from the study and can only be achieved by maintaining effective communication with all the scientific, technical and administrative personnel involved, and for a multi-site study with Principal Investigator(s). Of necessity, lines of communication should ensure that deviations from the study plan can be rapidly transmitted and that issues arising are documented.

If data are recorded on paper, the Study Director should ensure that the data generated are fully and accurately documented and that they have been generated in compliance with GLP Principles. For data recorded electronically onto a computerised system, the Study Director's responsibilities are the same as for paper systems. In addition, the Study Director should also ensure that computerised systems are suitable for their intended purpose, have been validated, and are fit for use in the study.

Final report

The final report of a study should be produced as a detailed scientific document outlining the purpose of the study, describing the methods and materials used, summarising and analysing data generated, and stating the conclusions drawn.

If the Study Director is satisfied that the report is a complete, true and accurate representation of the study and its results, then and only then, should the Study Director sign and date the final report to indicate acceptance of responsibility for the validity of the data. The extent of compliance with the GLP Principles should be indicated. He should also assure himself that there is a QA statement and that any deviations from the study plan have been noted.

Archives

On completion (including termination) of a study, the Study Director is responsible for ensuring that the study plan, final report, raw data and related material are archived in a timely manner. The final report should include a statement indicating where all the samples of test and reference items, specimens, raw data, study plan, final report and other related documentation are to be stored. Once data are transferred to the archives, the responsibility for it lies with management.

Sub-contracting

Where parts of any study are contracted out, the Study Director (and QA staff) should have knowledge of the GLP compliance status of that facility. If a contract facility is not GLP compliant, the Study Director must indicate this in the final report.

Study plan amendments and deviations

Study plan amendment

A study plan amendment should be issued to document an intended change in study design after the study initiation date and before the event occurs. An amendment may also be issued as a result of unexpected occurrences during the study that will require significant action. Amendments should indicate the reason for the change and be sequentially numbered, dated, signed and distributed to all recipients of the original study plan by the Study Director.

Study deviations

Whereas an amendment is an intended change to the study plan, a deviation is an unintended change which occurs during the execution of the study. Study information such as a deviation from the study plan should be noted in the study documentation. Such notes may be initiated by other personnel involved in the study, but should be acknowledged, described, explained and dated in a timely fashion by the Study Director and/or Principal Investigator(s) and maintained with the study raw data. The Study Director should approve any corrective action taken. The Study Director should consider whether to consult with other scientists to determine the impact of any such information on the study, and should report (and discuss where necessary) these deviations in the final report.

Qualifications of the Study Director

Qualifications for a Study Director will be dictated by the requirements of each individual study. Setting the criteria is the responsibility of the management. Furthermore, management has the responsibility for selection, monitoring and support of the Study Director to ensure that studies are carried out in compliance with the GLP Principles. Any minimal qualifications established by management for the position of Study Director should be documented in the appropriate personnel records. In addition to a strong technical background, the coordinating role of the Study Director requires an individual with strengths in communication and problem-solving and managerial skills.

Interface with the study

The Study Director has the overall responsibility for the conduct of a study. The term "responsibility for the overall conduct of the study and for its final report" may be interpreted in a broad sense for those studies where the Study Director may be geographically remote from parts of the actual experimental work. With multiple levels of management, study personnel and QA staff, it is critical that there are clear lines of authority and communication, and assigned responsibilities, so that the Study Director can effectively carry out his GLP responsibilities. This should be documented in writing. Test facility management should ensure that, for multi-site studies, clear lines of communication exist between the Study Director, Principal Investigator(s), the Quality Assurance Programme(s) and the study personnel

For studies that have delegated responsibilities to a Principal Investigator(s), the Study Director will rely on that individual to assure that relevant phase(s) of the study are conducted in accordance with the study plan, relevant SOPs and with GLP Principles. The Principal Investigator should contact the Study Director when event(s) occur that may affect the objectives defined in the study plan. All communications should be documented.

Communication between the Study Director and QA is required at all stages of the study.

This communication may involve:

- An active involvement with QA, for example, review of study plans in a timely manner, involvement in the review of new and revised Standard Operating Procedures, attendance of QA personnel at study initiation meetings and in resolving potential problems related to GLP.
- Responding to inspection and audit reports promptly, indicating corrective action and, if necessary, liaising with QA staff and scientific and technical personnel to facilitate responses to inspection/audit findings.

Replacement of the Study Director

The Study Director has the responsibility for the overall conduct of a study according to the GLP Principles and he has to ascertain that, in every phase of a study, these principles are fully complied with, that the study plan is followed faithfully and that all observations are fully documented. Theoretically, this responsibility can only be fulfilled if the Study Director is present all the time during the whole study. This is not always feasible in practice and there will be periods of absence which might make replacement necessary. While the circumstances under which a Study Director would be replaced are not defined in the GLP Principles, they should be addressed to the degree feasible by the facility's SOPs. These SOPs should also address the procedures and documentation necessary to replace a Study Director.

The decision for replacement or temporary delegation is the responsibility of management. All such decisions should be documented in writing. There are two circumstances where replacement might be considered, both of which are of importance only in longer-term studies, since the continuing presence of a Study Director during a short study may be assumed. In the event of termination of employment of a Study Director, the need for replacing this key person is obvious. In this case, one of the responsibilities of the replacement Study Director is, with the assistance of Quality Assurance personnel, to assure himself as soon as practicable of the GLP compliance in the study as conducted to date. The replacement of a Study Director and the reasons for it must be documented and authorised by management. It is also recommended that the results of any interim GLP review should be documented in case deficiencies or deviations have been found.

The second circumstance is when a Study Director is temporarily absent because of holidays, scientific meeting, illness or accident. An absence of short duration might not necessitate the formal replacement of the Study Director if it is possible to communicate with him if problems or emergencies arise. If critical study phases are expected to fall into the period of absence, they may either be moved to a more suitable time (with study plan amendment, if necessary), or a replacement of the Study Director may be considered, either by formally nominating a replacement Study Director or by temporary delegation of responsibilities to competent staff for this specific phase of the study. Should the unavailability of the Study Director be of longer duration, a replacement should be named rather than delegation to competent staff.

The returning Study Director must ascertain as soon as practicable whether or not deviations from GLP Principles have occurred, irrespective of whether or not he was formally replaced during his absence. Deviations from GLP Principles during his absence should be documented by the returning Study Director.

Legal status of the Study Director

The Study Director, by virtue of his signature in the final report confirming compliance with the GLP Principles, assumes responsibility for the performance of the study in compliance with GLP Principles and for the accurate representation of the raw data in the final report. However, the legal liability of the Study Director is established by national legislation and legal processes, and not by the OECD Principles of GLP.

6. THE APPLICATION OF THE PRINCIPLES OF GLP TO COMPUTERISED SYSTEMS*

History

Within the framework of the Third OECD Consensus Workshop on Good Laboratory Practice held 5-8 October 1992, in Interlaken, Switzerland, a working group of experts discussed the interpretation of the GLP Principles as applied to computerised systems. The working group was chaired by Dr. Theo Helder of the Dutch GLP Compliance Monitoring Authority. The Rapporteur was Mr. Bryan Doherty (Chairman of the Computing Committee of the British Association for Research Quality Assurance). Participants in the working group were from both national GLP compliance monitoring authorities and from testing laboratories in the following countries: Austria, Belgium, Denmark, Finland, France, Germany, Japan, the Netherlands, Switzerland, the United Kingdom, the United States. That working group was unable to reach consensus on a detailed guidance document in the time available to it. It did, however, develop a document entitled "Concepts relating to Computerised Systems in a GLP Environment", which set out the general principles and described the issues involved for each. That document was circulated to comments to member countries.

In light of the comments received, the Panel on Good Laboratory Practice, at its Fifth Meeting in March 1993, agreed that further work needed to be done and called for a second working group meeting to be held. Under the chairmanship of Dr. Helder, and with Mr. Doherty as rapporteur, that group met in Paris from 14 to 16 December 1994. Participants representing government and industry from Canada, Denmark, France, Germany, Japan, the Netherlands, Sweden, the United Kingdom and the United States took part.

The draft Consensus Document developed by the working group was based on the document emanating from the Interlaken workshop, comments from member countries thereto and a document developed by a United Kingdom joint government-industry working party. It was subsequently reviewed, modified and endorsed by the Panel and the Joint Meeting of the Chemicals Group and Management Committee of the Special Programme on the Control of Chemicals in 1995. It is declassified under the authority of the Secretary-General.

Introduction

Throughout recent years there has been an increase in the use of computerised systems by test facilities undertaking health and environmental safety testing. These computerised systems may be involved with the direct or indirect capture of data, processing, reporting and storage of data, and increasingly as an integral part of automated equipment. Where

---

* No. 10 in the OECD Series on Principles of Good Laboratory Practice and Compliance Monitoring.

these computerised systems are associated with the conduct of studies intended for regulatory purposes, it is essential that they are developed, validated, operated and maintained in accordance with the OECD Principles of Good Laboratory Practice (GLP).

Scope

All computerised systems used for the generation, measurement or assessment of data intended for regulatory submission should be developed, validated, operated and maintained in ways which are compliant with the GLP Principles.

During the planning, conduct and reporting of studies there may be several computerised systems in use for a variety of purposes. Such purposes might include the direct or indirect capture of data from automated instruments, operation/control of automated equipment and the processing, reporting and storage of data. For these different activities, computerised systems can vary from a programmable analytical instrument, or a personal computer to a laboratory information management system (LIMS) – with multiple functions. Whatever the scale of computer involvement, the GLP Principles should be applied.

Approach

Computerised systems associated with the conduct of studies destined for regulatory submission should be of appropriate design, adequate capacity and suitable for their intended purposes. There should be appropriate procedures to control and maintain these systems, and the systems should be developed, validated and operated in a way which is in compliance with the GLP Principles.

The demonstration that a computerised system is suitable for its intended purpose is of fundamental importance and is referred to as computer validation.

The validation process provides a high degree of assurance that a computerised system meets its pre-determined specifications. Validation should be undertaken by means of a formal validation plan and performed prior to operational use.

The application of the GLP Principles to computerised systems

The following considerations will assist in the application of the GLP Principles to computerised systems outlined above:

1. *Responsibilities*

    a) *Management* of a test facility has the overall responsibility for compliance with the GLP Principles. This responsibility includes the appointment and effective organisation of an adequate number of appropriately qualified and experienced staff, as well as the obligation to ensure that the facilities, equipment and data handling procedures are of an adequate standard.

    Management is responsible for ensuring that computerised systems are suitable for their intended purposes. It should establish computing policies and procedures to ensure that systems are developed, validated, operated and maintained in accordance with the GLP Principles. Management should also ensure that these policies and procedures are understood and followed, and ensure that effective monitoring of such requirements occurs.

Management should also designate personnel with specific responsibility for the development, validation, operation and maintenance of computerised systems. Such personnel should be suitably qualified, with relevant experience and appropriate training to perform their duties in accordance with the GLP Principles.

b) *Study Directors* are responsible under the GLP Principles for the overall conduct of their studies. Since many such studies will utilise computerised systems, it is essential that Study Directors are fully aware of the involvement of any computerised systems used in studies under their direction.

The Study Director's responsibility for data recorded electronically is the same as that for data recorded on paper and thus only systems that have been validated should be used in GLP studies.

c) *Personnel*. All personnel using computerised systems have a responsibility for operating these systems in compliance with the GLP Principles. Personnel who develop, validate, operate and maintain computerised systems are responsible for performing such activities in accordance with the GLP Principles and recognized technical standards.

d) *Quality Assurance* (QA) responsibilities for computerised systems must be defined by management and described in written policies and procedures. The Quality Assurance Programme should include procedures and practices that will assure that established standards are met for all phases of the validation, operation and maintenance of computerised systems. It should also include procedures and practices for the introduction of purchased systems and for the process of in-house development of computerised systems.

Quality Assurance personnel are required to monitor the GLP compliance of computerised systems and should be given training in any specialist techniques necessary. They should be sufficiently familiar with such systems so as to permit objective comment; in some cases, the appointment of specialist auditors may be necessary.

QA personnel should have, for review, direct read-only access to the data stored within a computerised system.

2. Training

The GLP Principles require that a test facility has appropriately qualified and experienced personnel and that there are documented training programmes including both on-the-job training and, where appropriate, attendance at external training courses. Records of all such training should be maintained.

The above provisions should also apply for all personnel involved with computerised systems.

3. Facilities and equipment

Adequate facilities and equipment should be available for the proper conduct of studies in compliance with GLP. For computerised systems there will be a number of specific considerations:

a) *Facilities*

Due consideration should be given to the physical location of computer hardware, peripheral components, communications equipment and electronic storage media. Extremes of temperature and humidity, dust, electromagnetic interference and proximity to high voltage cables should be avoided unless the equipment is specifically designed to operate under such conditions.

Consideration must also be given to the electrical supply for computer equipment and, where appropriate, back-up or uninterruptable supplies for computerised systems, whose sudden failure would affect the results of a study.

Adequate facilities should be provided for the secure retention of electronic storage media.

b) *Equipment*

i) *Hardware and software*

A computerised system is defined as a group of hardware components and associated software designed and assembled to perform a specific function or group of functions.

Hardware is the physical components of the computerised system; it will include the computer unit itself and its peripheral components.

Software is the programme or programmes that control the operation of the computerised system.

All GLP Principles which apply to equipment therefore apply to both hardware and software.

ii) *Communications*

Communications related to computerised systems broadly fall into two categories: between computers or between computers and peripheral components.

All communication links are potential sources of error and may result in the loss or corruption of data. Appropriate controls for security and system integrity must be adequately addressed during the development, validation, operation and maintenance of any computerised system.

4. **Maintenance and disaster recovery**

All computerised systems should be installed and maintained in a manner to ensure the continuity of accurate performance.

a) *Maintenance*

There should be documented procedures covering both routine preventative maintenance and fault repair. These procedures should clearly detail the roles and responsibilities of personnel involved. Where such maintenance activities have necessitated changes to hardware and/or software, it may be necessary to validate the system again. During the daily operation of the system, records should be maintained of any problems or inconsistencies detected and any remedial action taken.

b) *Disaster recovery*

Procedures should be in place describing the measures to be taken in the event of partial or total failure of a computerised system. Measures may range from planned hardware redundancy to transition back to a paper-based system. All contingency plans need to be well documented, validated and should ensure continued data integrity and should not compromise the study in any way. Personnel involved in the conduct of studies according to the GLP Principles should be aware of such contingency plans.

Procedures for the recovery of a computerised system will depend on the criticality of the system, but it is essential that back-up copies of all software are maintained. If recovery procedures entail changes to hardware or software, it may be necessary to validate the system again.

5. Data

The GLP Principles define raw data as being all original laboratory records and documentation, including data directly entered into a computer through an instrument interface, which are the results of original observations and activities in a study and which are necessary for the reconstruction and evaluation of the report of that study.

Computerised systems operating in compliance with GLP Principles may be associated with raw data in a variety of forms, for example, electronic storage media, computer or instrument printouts and microfilm/fiche copies. It is necessary that raw data are defined for each computerised system.

Where computerised systems are used to capture, process, report or store raw data electronically, system design should always provide for the retention of full audit trails to show all changes to the data without obscuring the original data. It should be possible to associate all changes to data with the persons making those changes by use of timed and dated (electronic) signatures. Reasons for change should be given.

When raw data are held electronically, it is necessary to provide for long term retention requirements for the type of data held and the expected life of computerised systems. Hardware and software system changes must provide for continued access to and retention of the raw data without integrity risks.

Supporting information such as maintenance logs and calibration records that are necessary to verify the validity of raw data or to permit reconstruction of a process or a study should be retained in the archives.

Procedures for the operation of a computerised system should also describe the alternative data capture procedures to be followed in the event of system failure. In such circumstances, any manually recorded raw data subsequently entered into the computer should be clearly identified as such, and should be retained as the original record. Manual back-up procedures should serve to minimise the risk of any data loss and ensure that these alternative records are retained.

Where system obsolescence forces a need to transfer electronic raw data from one system to another, then the process must be well documented and its integrity verified. Where such migration is not practicable, then the raw data must be transferred to another medium and this verified as an exact copy prior to any destruction of the original electronic records.

6. Security

Documented security procedures should be in place for the protection of hardware, software and data from corruption or unauthorised modification, or loss. In this context, security includes the prevention of unauthorised access or changes to the computerised system as well as to the data held within the system. The potential for corruption of data by viruses or other agents should also be addressed. Security measures should also be taken to ensure data integrity in the event of both short term and long term system failure.

a) Physical security

Physical security measures should be in place to restrict access to computer hardware, communications equipment, peripheral components and electronic storage media to authorised personnel only. For equipment not held within specific "computer rooms" (*e.g.* personal computers and terminals), standard test facility access controls are necessary as a minimum. However, where such equipment is located remotely (*e.g.* portable components and modem links), additional measures need to be taken.

b) Logical security

For each computerised system or application, logical security measures must be in place to prevent unauthorised access to the computerised system, applications and data. It is essential to ensure that only approved versions and validated software are in use. Logical security may include the need to enter a unique user identity with an associated password. Any introduction of data or software from external sources should be controlled. These controls may be provided by the computer operating system software, by specific security routines, routines embedded into the applications or combinations of the above.

c) Data integrity

Since maintaining data integrity is a primary objective of the GLP Principles, it is important that everyone associated with a computerised system is aware of the necessity for the above security considerations. Management should ensure that personnel are aware of the importance of data security, the procedures and system features that are available to provide appropriate security and the consequences of security breaches. Such system features could include routine surveillance of system access, the implementation of file verification routines and exception and/or trend reporting.

d) Back-up

It is standard practice with computerised systems to make back-up copies of all software and data to allow for recovery of the system following any failure which compromises the integrity of the system, *e.g.* disk corruption. The implication, therefore, is that the back-up copy may become raw data and must be treated as such.

7. Validation of computerised systems

Computerised systems must be suitable for their intended purpose. The following aspects should be addressed:

a) Acceptance

Computerised systems should be designed to satisfy GLP Principles and introduced in a pre-planned manner. There should be adequate documentation that each system was developed in a controlled manner and preferably according to recognised quality and technical standards (*e.g.* ISO 9001). Furthermore, there should be evidence that the system was adequately tested for conformance with the acceptance criteria by the test facility prior to being put into routine use. Formal acceptance testing requires the conduct of tests following a pre-defined plan and retention of documented evidence of all testing procedures, test data, test results, a formal summary of testing and a record of formal acceptance.

For vendor-supplied systems, it is likely that much of the documentation created during the development is retained at the vendor's site. In this case, evidence of formal assessment and/or vendor audits should be available at the test facility.

b) *Retrospective evaluation*

There will be systems where the need for compliance with GLP Principles was not foreseen or not specified. Where this occurs, there should be documented justification for use of the systems; this should involve a retrospective evaluation to assess suitability.

Retrospective evaluation begins by gathering all historical records related to the computerised system. These records are then reviewed and a written summary is produced. This retrospective evaluation summary should specify what validation evidence is available and what needs to be done in the future to ensure validation of the computerised system.

c) *Change control*

Change control is the formal approval and documentation of any change to the computerised system during the operational life of the system. Change control is needed when a change may affect the computerised system's validation status. Change control procedures must be effective once the computerised system is operational.

The procedure should describe the method of evaluation to determine the extent of retesting necessary to maintain the validated state of the system. The change control procedure should identify the persons responsible for determining the necessity for change control and its approval. Irrespective of the origin of the change (supplier or in-house developed system), appropriate information needs to be provided as part of the change control process. Change control procedures should ensure data integrity.

d) *Support mechanism*

In order to ensure that a computerised system remains suitable for its intended purpose, support mechanisms should be in place to ensure the system is functioning and being used correctly. This may involve system management, training, maintenance, technical support, auditing and/or performance assessment. Performance assessment is the formal review of a system at periodic intervals to ensure that it continues to meet stated performance criteria, *e.g.* reliability, responsiveness, capacity.

8. **Documentation**

The items listed below are a guide to the minimum documentation for the development, validation, operation and maintenance of computerised systems.

a) *Policies*

There should be written management policies covering, *inter alia*, the acquisition, requirements, design, validation, testing, installation, operation, maintenance, staffing, control, auditing, monitoring and retirement of computerised systems.

b) *Application description*

For each application, there should be documentation fully describing:

- The name of the application software or identification code and a detailed and clear description of the purpose of the application.
- The hardware (with model numbers) on which the application software operates.
- The operating system and other system software (*e.g.* tools) used in conjunction with the application.
- The application programming language(s) and/or data base tools used.
- The major functions performed by the application
- An overview of the type and flow of data/data base design associated with the application. File structures, error and alarm messages, and algorithms associated with the application.
- The application software components with version numbers.
- Configuration and communication links among application modules and to equipment and other systems.

c) *Source code*

Some OECD member countries require that the source code for application software should be available at, or retrievable to, the test facility.

d) *Standard Operating Procedures (SOPs)*

Much of the documentation covering the use of computerised systems will be in the form of SOPs. These should cover, but not be limited to, the following:

- Procedures for the operation of computerised systems (hardware/software), and the responsibilities of personnel involved.
- Procedures for security measures used to detect and prevent unauthorised access and programme changes.
- Procedures and authorisation for programme changes and the recording of changes.
- Procedures and authorisation for changes to equipment (hardware/software), including testing before use if appropriate.
- Procedures for the periodic testing for correct functioning of the complete system or its component parts and the recording of these tests.
- Procedures for the maintenance of computerised systems and any associated equipment.

- Procedures for software development and acceptance testing, and the recording of all acceptance testing.
- Back-up procedures for all stored data and contingency plans in the event of a breakdown. Procedures for the archiving and retrieval of all documents, software and computer data.
- Procedures for the monitoring and auditing of computerised systems.

9. Archives

The GLP Principles for archiving data must be applied consistently to all data types. It is therefore important that electronic data are stored with the same levels of access control, indexing and expedient retrieval as other types of data.

Where electronic data from more than one study are stored on a single storage medium (*e.g.* disk or tape), a detailed index will be required.

It may be necessary to provide facilities with specific environmental controls appropriate to ensure the integrity of the stored electronic data. If this necessitates additional archive facilities, then management should ensure that the personnel responsible for managing the archives are identified and that access is limited to authorised personnel. It will also be necessary to implement procedures to ensure that the long-term integrity of data stored electronically is not compromised. Where problems with long-term access to data are envisaged or when computerised systems have to be retired, procedures for ensuring that continued readability of the data should be established. This may, for example, include producing hard copy printouts or transferring the data to another system.

No electronically stored data should be destroyed without management authorization and relevant documentation. Other data held in support of computerised systems, such as source code and development, validation, operation, maintenance and monitoring records, should be held for at least as long as study records associated with these systems.

Definitions of terms

Acceptance criteria: The documented criteria that should be met to successfully complete a test phase or to meet delivery requirements.

Acceptance testing: Formal testing of a computerised system in its anticipated operating environment to determine whether all acceptance criteria of the test facility have been met and whether the system is acceptable for operational use.

Back-up: Provisions made for the recovery of data files or software, for the restart of processing, or for the use of alternative computer equipment after a system failure or disaster.

Change control: Ongoing evaluation and documentation of system operations and changes to determine whether a validation process is necessary following any changes to the computerised system. Computerised System: A group of hardware components and associated software designed and assembled to perform a specific function or group of functions.

Electronic signature: The entry in the form of magnetic impulses or computer data compilation of any symbol or series of symbols, executed, adapted or authorised by a person to be equivalent to the person's handwritten signature.

Hardware: The physical components of a computerised system, including the computer unit itself and its peripheral components.

Peripheral Components: Any interfaced instrumentation, or auxiliary or remote components such as printers, modems and terminals, etc.

Recognised technical standards: Standards as promulgated by national or international standard setting bodies (ISO, IEEE, ANSI, etc.).

Security: The protection of computer hardware and software from accidental or malicious access, use, modification, destruction or disclosure. Security also pertains to personnel, data, communications and the physical and logical protection of computer installations.

Software (application): A programme acquired for or developed, adapted or tailored to the test facility requirements for the purpose of controlling processes, data collection, data manipulation, data reporting and/or archiving.

Software (operating system): A programme or collection of programmes, routines and sub-routines that controls the operation of a computer. An operating system may provide services such as resource allocation, scheduling, input/output control, and data management.

Source code: An original computer programme expressed in human-readable form (programming language) which must be translated into machine-readable form before it can be executed by the computer.

Validation of a computerised system: The demonstration that a computerised system is suitable for its intended purpose.

Further definitions of terms can be found in the OECD Principles of Good Laboratory Practice (Part I of this publication).

7. THE APPLICATION OF THE OECD PRINCIPLES OF GLP TO THE ORGANISATION AND MANAGEMENT OF MULTI-SITE STUDIES*

History

It is becoming increasingly common for non-clinical health and environmental safety studies to be conducted at more than one site. For example, companies may use facilities which specialise in different activities located at sites in various countries; or field trials on agrochemicals may have to be conducted on different crops or soil types located in different regions or countries. Toxicology studies may also have phases of the study conducted by different departments of the same organisation or different companies.

In the framework of the Second OECD Consensus Workshop on Good Laboratory Practice, held 21-23 May 1991, in Vail, Colorado, experts discussed and reached consensus on the application of the GLP Principles to field studies. An OECD Consensus Document on "The Application of the GLP Principles to Field Studies" was subsequently published in 1992 and revised in 1999 (Part II, chapter 3 in this publication). Among other aspects, this document introduced the concept of a "Principal Investigator" who could assume delegated responsibility for a phase of a field study being conducted at a test site that was remote from the Study Director. Although the concept of a Principal Investigator had originally been developed to assist in the conduct of field studies that included trials being conducted at several different locations, the concept is equally applicable to any other type of multi-site study.

The Revised OECD Principles of Good Laboratory Practice published in 1997 now refer to the role of the Principal Investigator in the conduct of any multi-site study.

A study can be a "multi-site" study for a variety of reasons. A single site that undertakes a study may not have the technical expertise or capability to perform a particular task that is needed, so this work is performed at another site. A sponsor who has placed a study at a contract research organisation may request that certain study activities, such as bioanalysis, be contracted out to a specified laboratory or the sponsor may request that specimens be returned to them for analysis.

The purpose of this document is to provide guidance on the issues that are involved in the planning, performance, monitoring, recording, reporting and archiving of multi-site studies. It was developed by the Fourth OECD Consensus Workshop in Horley, United Kingdom, in June 2001. It was endorsed by the Working Group on GLP in December 2001 and, subsequently, by the Joint Meeting of the Chemicals Committee and Working Party on Chemicals, Pesticides and Biotechnology in February 2002. It is declassified under the authority of the Secretary-General.

---

* No. 13 in the OECD Series on Principles of Good Laboratory Practice and Compliance Monitoring.

Introduction

The planning, performance, monitoring, recording, reporting and archiving of a multi-site study present a number of potential problems that should be addressed to ensure that the GLP compliance of the study is not compromised. The fact that different study activities are being conducted at different sites means that the planning, communication and control of the study are of vital importance.

Although a multi-site study will consist of work being conducted at more than one site (which includes the test facility and all test sites), it is still a single study that should be conducted in accordance with the OECD Principles of GLP. This means that there should be a single study plan, a single Study Director, and ultimately, a single final report. It is therefore essential that, when the study is first planned, personnel and management at the contributing sites are made aware that the work they will perform is part of a study under the control of the Study Director and is not to be carried out as a separate study.

It is imperative that the work to be carried out by the various sites is clearly identified at an early stage of planning, so that the necessary control measures can be agreed upon by the parties concerned before the study plan is finalised.

Many of the problems associated with the conduct of multi-site studies can be prevented by clear allocation of responsibilities and effective communication among all parties involved in the conduct of the study. This will include the sponsor, the Study Director, and the management, the Principal Investigator(s), Quality Assurance and study personnel at each site.

All of these parties should be aware that, when a multi-site study is conducted in more than one country, there might be additional issues due to differences in national culture, language and GLP compliance monitoring programmes. In these situations, it may be necessary to seek the advice of the national GLP compliance monitoring authority where the site is located.

The guidance contained within this document should be considered during the planning, performance, monitoring, recording, reporting and archiving of any study that will be conducted at more than one site. The guidance applies to all types of non-clinical health and environmental safety studies.

Management and control of multi-site studies

A *multi-site study* means any study that has phases conducted at more than one site. Multi-site studies become necessary if there is a need to use sites that are geographically remote, organisationally distinct or otherwise separated. This could include a department of an organisation acting as a test site when another department of the same organisation acts as the test facility.

A *phase* is a defined activity or set of activities in the conduct of a study.

The decision to conduct a multi-site study should be carefully considered by the sponsor in consultation with test facility management assigned by the sponsor before study initiation. The use of multiple test sites increases the complexity of study design and management tasks, resulting in additional risks to study integrity. It is therefore important that all of the potential threats to study integrity presented by a multi-site configuration are evaluated, that responsibilities are clear and that risks are minimised. Full consideration should be given to the technical/scientific expertise, GLP compliance status, resources and commercial viability of all of the test sites that may be used.

Communication

For a multi-site study to be conducted successfully, it is imperative that all parties involved are aware of their responsibilities. In order to discharge these responsibilities, and to deal with any events that may need to be addressed during the conduct of the study, the flow of information and effective communication among the sponsor, management at sites, the Study Director, Principal Investigator(s), Quality Assurance and study personnel is of paramount importance.

The mechanism for communication of study-related information among these parties should be agreed in advance and documented.

The Study Director should be kept informed of the progress of the study at all sites.

Study management

The sponsor will assign a study to a test facility. Test facility management will appoint the Study Director who need not necessarily be located at the site where the majority of the experimental work is done. The decision to conduct study activities at other sites will usually be made by test facility management in consultation with the Study Director and the sponsor, where necessary.

When the Study Director is unable to perform his/her duties at a test site because of geographical or organisational separation, the need to appoint a Principal Investigator(s) at a test site(s) arises. The performance of duties may be impracticable, for example, because of travel time, time zones, or delays in language interpretation. Geographical separation may relate to distance or to the need for simultaneous attention at more than one location.

Test facility management should facilitate good working relationships with test site management to ensure study integrity. The preferences of the different groups involved, or commercial and confidentiality agreements, should not preclude the exchange of information necessary to ensure proper study conduct.

Roles and responsibilities

Sponsor

The decision to conduct a multi-site study should be carefully considered by the sponsor in consultation with test facility management before study initiation. The sponsor should specify whether compliance with the OECD Principles of GLP and applicable national legislation is required. The sponsor should understand that a multi-site study must result in one final report.

The sponsor should be aware that, if its site acts as a test site undertaking a phase(s) of a multi-site study, its operations and staff involved in the study are subject to control of the Study Director. According to the specific situation, this may include visits from test facility management, the Study Director and/or inspections by the lead Quality Assurance. The Study Director has to indicate the extent to which the study complies with GLP, including any work conducted by the sponsor.

Test facility management

Test facility management should approve the selection of test sites. Issues to consider will include, but are not limited to, practicality of communication, adequacy of Quality Assurance arrangements, and the availability of appropriate equipment and expertise. Test facility management should designate a lead Quality Assurance that has the overall responsibility for quality assurance of the entire study. Test facility management should inform all test site quality assurance units of the location of the lead Quality Assurance. If it is necessary to use a test site that is not included in a national GLP compliance monitoring programme, the rationale for selection of this test site should be documented. Test facility management should make test site management aware that it may be subject to inspection by the national GLP compliance monitoring authority of the country in which the test site is located. If there is no national GLP compliance monitoring authority in that country, the test site may be subject to inspection by the GLP compliance monitoring authority from the country to which the study has been submitted.

Test site management

Test site management is responsible for the provision of adequate site resources and for selection of appropriately skilled Principal Investigator(s). If it becomes necessary to replace a Principal Investigator, test site management will appoint a replacement Principal Investigator in consultation with the sponsor, the Study Director and test facility management where necessary. Details should be provided to the Study Director in a timely manner so that a study plan amendment can be issued. The replacement Principal Investigator should assess the GLP compliance status of the work conducted up to the time of replacement.

Study Director

The Study Director should ensure that the test sites selected are acceptable. This may involve visits to test sites and meetings with test site personnel.

If the Study Director considers that the work to be done at one of the test sites can be adequately controlled directly by him(her)self without the need for a Principal Investigator to be appointed, he/she should advise test facility management of this possibility. Test facility management should ensure that appropriate quality assurance monitoring of that site is arranged. This could be by the test site's own Quality Assurance or by the lead Quality Assurance.

The Study Director is responsible for the approval of the study plan, including the incorporation of contributions from Principal Investigators. The Study Director will approve and issue amendments to and acknowledge deviations from the study plan, including those relating to work undertaken at sites. The Study Director is responsible for ensuring that all staff are clearly aware of the requirements of the study and should ensure that the study plan and amendments are available to all relevant personnel.

The Study Director should set up, test and maintain appropriate communication systems between him(her)self and each Principal Investigator. For example, it is prudent to verify telephone numbers and electronic mail addresses by test transmissions, to consider signal strength at rural field stations, etc. Differences in time zones may need to be taken into account. The Study Director should liaise directly with each Principal Investigator and not via an intermediary, except where this is unavoidable (*e.g.* the need for language interpreters).

Throughout the conduct of the study, the Study Director should be readily available to the Principal Investigators. The Study Director should facilitate the co-ordination and timing of events and movement of samples, specimens or data between sites, and ensure that Principal Investigators understand chain of custody procedures.

The Study Director should liaise with Principal Investigators about test site quality assurance findings as necessary. All communication between the Study Director and Principal Investigators or test site quality assurance in relation to these findings should be documented.

The Study Director should ensure that the final report is prepared, incorporating any contributions from Principal Investigators. The Study Director should ensure that the final report is submitted to the lead Quality Assurance for inspection. The Study Director will sign and date the final report to indicate the acceptance of responsibility for the validity of the data and to indicate the extent to which the study complies with the OECD Principles of Good Laboratory Practice. This may be based partly on written assurances provided by the Principal Investigator(s).

At sites where no Principal Investigator has been appointed, the Study Director should liaise directly with the personnel conducting the work at those sites. These personnel should be identified in the study plan.

Principal Investigator

The Principal Investigator acts on behalf of the Study Director for the delegated phase and is responsible for ensuring compliance with the Principles of GLP for that phase. A fully co-operative, open working relationship between the Principal Investigator and the Study Director is essential.

There should be documented agreement that the Principal Investigator will conduct the delegated phase in accordance with the study plan and the Principles of GLP. Signature of the study plan by the Principal Investigator would constitute acceptable documentation.

Deviations from the study plan or Standard Operating Procedures (SOPs) related to the study should be documented at the test site, be acknowledged by the Principal Investigator and reported to and acknowledged by the Study Director in a timely manner.

The Principal Investigator should provide the Study Director with contributions which enable the preparation of the final report. These contributions should include written assurance from the Principal Investigator confirming the GLP compliance of the work for which he/she is responsible.

The Principal Investigator should ensure that all data and specimens for which he/she is responsible are transferred to the Study Director or archived as described in the study plan. If these are not transferred to the Study Director, the Principal Investigator should notify the Study Director when and where they have been archived. During the study, the Principal Investigator should not dispose of any specimens without the prior written permission of the Study Director.

Study personnel

The GLP Principles require that all professional and technical personnel involved in the conduct of a study have a job description and a record of the training, qualifications and experience which support their ability to undertake the tasks assigned to them. Where study personnel are required to follow approved SOPs from another test site, any additional training required should be documented.

There may be some sites where temporarily employed personnel carry out aspects of study conduct. Where these persons have generated or entered raw data, or have performed activities relevant to the conduct of the study, records of their qualifications, training and experience should be maintained. Where these individuals have carried out routine operations such as livestock handling subject to supervision by more highly qualified staff, no such personnel records need be maintained.

Quality assurance

The quality assurance of multi-site studies needs to be carefully planned and organised to ensure that the overall GLP compliance of the study is assured. Because there is more than one site, issues may arise with multiple management organisations and Quality Assurance programmes.

Responsibilities of lead Quality Assurance

The lead Quality Assurance should liaise with test site quality assurance to ensure adequate quality assurance inspection coverage throughout the study.

Particular attention should be paid to the operation and documentation relating to communication among sites. Responsibilities for quality assurance activities at the various sites should be established before experimental work commences at those sites.

The lead Quality Assurance will ensure that the study plan is verified and that the final report is inspected for compliance with the Principles of GLP. Quality assurance inspections of the final report should include verification that the Principal Investigator contributions (including evidence of quality assurance at the test site) have been properly incorporated. The lead Quality Assurance will ensure that a Quality Assurance statement is prepared relating to the work undertaken by the test facility including or referencing quality assurance statements from all test sites.

Responsibilities of test site quality assurance

Each test site management is usually responsible for ensuring that there is appropriate quality assurance for the part of the study conducted at their site. Quality assurance at each test site should review sections of the study plan relating to operations to be conducted at their site. They should maintain a copy of the approved study plan and study plan amendments.

Quality assurance at the test site should inspect study-related work at their site according to their own SOPs, unless required to do otherwise by the lead Quality Assurance, reporting any inspection results promptly in writing to the Principal Investigator, test site management, Study Director, test facility management and lead Quality Assurance.

Quality assurance at the test site should inspect the Principal Investigator's contribution to the study according to their own test site SOPs and provide a statement relating to the quality assurance activities at the test site.

Master schedules

A multi-site study in which one or more Principal Investigators have been appointed should feature on the master schedule of all sites concerned. It is the responsibility of test facility management and test site management to ensure that this is done.

The unique identification of the study must appear on the master schedule in each site, cross-referenced as necessary to test site identifiers. The Study Director should be identified on the master schedule(s), and the relevant Principal Investigator shown on each site master schedule.

At all sites, the start and completion dates of the study phase(s) for which they are responsible should appear on their master schedule.

Study plan

For each multi-site study, a single study plan should be issued. The study plan should clearly identify the names and addresses of all sites involved.

The study plan should include the name and address of any Principal Investigators and the phase of the study delegated to them. It is recommended that sufficient information is included to permit direct contact by the Study Director, *e.g.* telephone number.

The study plan should identify how data generated at sites will be provided to the Study Director for inclusion in the final report.

It is useful, if known, to describe in the study plan the location(s) at which the data, samples of test and reference items and specimens generated at the different sites are to be retained.

It is recommended that the draft study plan should be made available to Principal Investigators for consideration and acknowledgement of their capability to undertake the work assigned to them, and to enable them to make any specialised technical contribution to the study plan if required.

The study plan is normally written in a single language, usually that of the Study Director. For multi-national studies, it may be necessary for the study plan to be issued in more than one language; this intention should be indicated in the original study plan, the translated study plan(s) and the original language should be identified in all versions. There will need to be a mechanism to verify the accuracy and completeness of the translated study plan. The responsibility for the accuracy of the translation can be delegated by the Study Director to a language expert and should be documented.

Performance of the study

This section repeats the most important requirements from the Principles of GLP and recommendations from the Consensus Document on the Application of the GLP Principles to Field Studies in order to provide useful guidance for organisation of multi-site studies. These documents should be consulted for further details.

Facilities

Sites may not have a full time staff presence during the working day. In this situation, it may be necessary to take additional measures to maintain the physical security of the test item, specimens and data.

When it is necessary to transfer data or any materials among sites, mechanisms to maintain their integrity need to be established. Special care needs to be taken when transferring data electronically (e-mail, Internet, etc.).

Equipment

Equipment being used in a study should be fit for its intended purpose. This is also applicable to large mechanical vehicles or highly specialised equipment that may be used at some sites.

There should be maintenance and calibration records for such equipment that serve to indicate their "fitness for intended purpose" at the time of use. Some apparatus (*e.g.* leased or rented equipment such as large animal scales and analytical equipment) may not have records of periodic inspection, cleaning, maintenance and calibration. In such cases, information should be recorded in the study-specific raw data to demonstrate "fitness for intended purpose" of the equipment.

Control and accountability of study material

Procedures should be in place that will ensure timely delivery of study-related materials to sites. Maintaining integrity/stability during transport is essential, so the use of reliable means of transportation and chain of custody documentation is critical. Clearly defined procedures for transportation, and responsibilities for who does what, are essential.

Adequate documentation should accompany each shipment of study material to satisfy any applicable legal requirements, *e.g.* customs, health and safety legislation. This documentation should also provide relevant information sufficient to ensure that it is suitable for its intended purpose on arrival at any site. These aspects should be resolved prior to shipment.

When study materials are transported between sites in the same consignment, it is essential that there is adequate separation and identification to avoid mix-ups or cross contamination. This is of particular importance if materials from more than one study are transported together.

If the materials being transported might be adversely affected by environmental conditions encountered during transportation, procedures should be established to preserve their integrity. It may be appropriate for monitoring to be carried out to confirm that required conditions were maintained.

Attention should be given to the storage, return or disposal of excess test and reference items being used at sites.

Reporting of study results

A single final report should be issued for each multi-site study. The final report should include data from all phases of the study. It may be useful for the Principal Investigators to produce a signed and dated report of the phase delegated to them, for incorporation into the final report. If prepared, such reports should include evidence that appropriate quality assurance monitoring was performed at that test site and contain sufficient commentary to enable the Study Director to write a valid final report covering the whole study. Alternatively, raw data may be transferred from the Principal Investigator to the Study Director, who should ensure that the data are presented in the final report. The final report produced in this way should identify the Principal Investigator(s) and the phase(s) for which they were responsible.

The Principal Investigators should indicate the extent to which the work for which they were responsible complies with the GLP Principles, and provide evidence of the quality assurance inspections performed at that test site. This may be incorporated directly into the final report, or the required details may be extracted and included in the Study Director's compliance claim and Quality Assurance statement in the final report. When details have been extracted, the source should be referenced and retained.

The Study Director must sign and date the final report to indicate acceptance of responsibility for the validity of all the data. The extent of compliance with the GLP Principles should be indicated with specific reference to the OECD Principles of GLP and Regulations with which compliance is being claimed. This claim of compliance will cover all phases of the study and should be consistent with the information presented in the Principal Investigator claims. Any sites not compliant with the OECD Principles of GLP should be indicated in the final report.

The final report should identify the storage location(s) of the study plan, samples of test and reference items, specimens, raw data and the final report. Reports produced by Principal Investigators should provide information concerning the retention of materials for which they were responsible.

Amendments to the final report may only be produced by the Study Director. Where the necessary amendment relates to a phase conducted at any test site, the Study Director should contact the Principal Investigator to agree appropriate corrective actions. These corrective actions must be fully documented.

If a Principal Investigator prepares a report, that report should, where appropriate, comply with the same requirements that apply to the final report.

Standard Operating Procedures (SOPs)

The GLP Principles require that appropriate and technically valid SOPs are established and followed. The following examples are procedures specific to multi-site studies:

- selection and monitoring of test sites;
- appointment and replacement of Principal Investigators;
- transfer of data, specimens and samples between sites;
- verification or approval of foreign language translations of study plans or SOPs; and
- storage, return or disposal of test and reference items being used at remote test sites.

The Principles of GLP require that SOPs should be immediately available to study personnel when they are conducting activities, regardless of where they are carrying out the work.

It is recommended that test site personnel should follow test site SOPs. When they are required to follow other procedures specified by the Study Director, for example SOPs provided by the test facility management, this requirement should be identified in the study plan. The Principal Investigator is responsible for ensuring that test site personnel are aware of the procedures to be followed and have access to the appropriate documentation.

If personnel at a test site are required to follow SOPs provided by the test facility management, it is necessary for test site management to give written acceptance.

When SOPs from a test facility have been issued for use at a test site, test facility management should ensure that any subsequent SOP revisions produced during the course of the study are also sent to the test site and the superseded versions are removed from use. The Principal Investigator should ensure that all test site personnel are aware of the revision and only have access to the current version.

When SOPs from a test facility are to be followed at test sites, it may be necessary for the SOPs to be translated into other languages. In this situation, it is essential that any translations be thoroughly checked to ensure that the instructions and meaning of the different language versions remain identical. The original language should be defined in the translated SOPs.

Storage and retention of records and materials

During the conduct of multi-site studies, attention should be given to the temporary storage of materials. Such storage facilities should be secure and protect the integrity of their contents. When data are stored away from the test facility, assurance will be needed of the site's ability to readily retrieve data which may be needed for review.

Records and materials need to be stored in a manner that complies with GLP Principles. When test site storage facilities are not adequate to satisfy GLP requirements, records and materials should be transferred to a GLP compliant archive.

Test site management should ensure that adequate records are available to demonstrate test site involvement in the study.

PART II

Chapter 2

Advisory Documents of the Working Group on GLP

8. THE ROLE AND RESPONSIBILITIES OF THE SPONSOR IN THE APPLICATION OF THE PRINCIPLES OF GOOD LABORATORY PRACTICE*

History

In the framework of the Revision of the OECD Principles on Good Laboratory Practice, the Expert Group was not able to reach consensus on whether and how to deal with the role and responsibilities of the sponsor of chemical safety studies in the Principles. The revised Principles of GLP contain several explicit references to the sponsor, and the issue is implicit in several other principles. However, there was no agreement on the need for and content of a separate section in the Principles on this matter.

On the recommendation of the Chairman of the Expert Group, the Panel on GLP therefore agreed to develop a document which could advise the testing industry as far as possible on current practice in member countries and the interpretation of Panel of the GLP Principles related to this issue. At its ninth meeting in March 1997, the panel endorsed a document drafted by a Task Group on the role and responsibilities of the sponsor. The Task Group had met in Lisbon on 8 and 9 January 1997, was chaired by Theo Helder (Netherlands), and comprised Panel members or their representatives from Canada, Finland, France, Germany, Portugal, Sweden, and Switzerland.

The Joint Meeting of the Chemicals Group and Management Committee of the Special Programme on the Control of Chemicals endorsed the document in 1998. It is declassified under the authority of the Secretary-General.

Introduction

Although the revised Principles of Good Laboratory Practice only explicitly assign a few responsibilities to the sponsor of a study, the sponsor has other implicit responsibilities. These arise from the fact that the sponsor is often the party who initiates one or more studies and directly submits the results thereof to regulatory authorities. The sponsor must therefore assume an active role in confirming that all non-clinical health and environmental safety studies were conducted in compliance with GLP. Sponsors cannot rely solely on the assurances of test facilities they may have contracted to arrange or perform such studies. The guidance given below attempts to outline both the explicit and implicit responsibilities of a sponsor necessary to fulfil his obligations.

Definition

"Sponsor means an entity which commissions, supports and/or submits a non-clinical health and environmental safety study." (See Revised OECD Principles of GLP, para. 2.2, point No. 5.)

---

* No. 11 in the OECD Series on Principles of Good Laboratory Practice and Compliance Monitoring.

Note: Sponsor can include:

- an entity* who initiates and support, by provision of financial or other resources, non-clinical health and environmental safety studies;
- an entity who submits non-clinical health and environmental safety studies to regulatory authorities in support of a product registration or other application for which GLP compliance is required.

Responsibilities of the sponsor

The sponsor should understand the requirements of the Principles of Good Laboratory Practice, in particular those related to the responsibilities of the test facility management and the Study Director/Principal Investigator.

> *Note:* If parts of the study are contracted out to subcontractors by the sponsor, the sponsor should be aware that the responsibility for the whole study remains with the Study Director, including the validity of the raw data and the report.

When commissioning a non-clinical health and environmental safety study, the sponsor should ensure that the test facility is able to conduct the study in compliance with GLP and that it is aware that the study is to be performed under GLP.

> *Note:* There are various tools for assessing the ability of a test facility to conduct a study in compliance with GLP. It can be useful for the sponsor to monitor contracted laboratories prior to the initiation of as well as during the study in accordance with its nature, length and complexity to ensure that its facilities, equipment, SOPs and personnel are according to GLP. If the test facility is in the national GLP compliance monitoring programme, the national monitoring authority** may also be contacted to determine the current GLP compliance status of the test facility.

Where several studies are presented to a regulatory authority in a single package, the responsibility for the integrity of the assembled package of unaltered final reports lies with the sponsor. It is necessary that the sponsor ensures that adequate communication links exist between his representatives and all parties conducting a study, such as the Study Director, Quality Assurance unit and test facility management.

The sponsor is explicitly mentioned in several of the requirements of the Revised OECD Principles of GLP:

> **Characterisation of test item:** "In cases where the test item is supplied by the sponsor, there should be a mechanism, developed in co-operation between the sponsor and the test facility, to verify the identity of the test item subject to the study." (See Revised Principles, para. 6.2, point No. 3.)

> *Note:* This requirement has been added to the revised GLP Principles in order to ensure that there is no mix-up of test items.

> **Study plan:** "The study plan should also be approved by the test facility management and the sponsor if required by national regulation or legislation in the country where the study is being performed." (See Revised Principles, para. 8.1, point No. 1.)

---

* "Entity" may include an individual, partnership, corporation, association, scientific or academic establishment, government agency, or organisational unit thereof, or any other legally identifiable body.
** Sponsors should be aware that, notwithstanding any contractual requirements for confidentiality, national GLP monitoring authorities have access to all data produced by a GLP compliance facility.

Note: Some member countries require approval of study plans by sponsors due to legal considerations related to responsibility for validity of test data.

Content of the study plan: "The study plan should contain... information concerning the sponsor and the test facility... the name and address of the sponsor." [See Revised Principles, para. 8.2, point No. 2*a*).]

"The Study Plan should contain... (the) date of approval of the study plan by signature of the test facility management and sponsor, if required by national regulation or legislation in the country where the study is being performed." [See Revised Principles, para. 8.2, point No. 3*a*).]

Content of the final report: "The final report should include... information concerning the sponsor and the test facility... name and address of the sponsor." [See Revised Principles, para. 9.2, point No. 2*a*).]

Storage and retention of records and materials: "If a test facility or an archive contracting facility goes out of business and has no legal successor, the archive should be transferred to the archives of the sponsor(s) of the study(ies)." (See Revised Principles, para. 10.4.)

Note: In this case, the sponsor is expected to arrange for an archive for the appropriate storage and retrieval of study plans, raw data, specimens, samples of test and reference items and final reports in accordance with the Principles of GLP.

Other issues

Provision of chemical safety information

The sponsor should inform the test facility of any known potential risks of the test item to human health or the environment as well as any protective measures which should be taken by test facility staff.

Characterisation of the test item

The Revised OECD Principles of GLP include several requirements related to the characterisation of the test item [*e.g.* para. 6.2, point Nos. 1 and 2; para. 9.2, point No. 1*d*)]. These requirements call for careful identification of the test item and description of its characteristics. This characterisation is carried out either by the contracted test facility or by the sponsor. If the characterisation is indeed conducted by the sponsor, this fact should be explicitly mentioned in the final report. Sponsors should be aware that failure to conduct characterisation in accordance with GLP case could lead to rejection of a study by a regulatory authority in some member countries.

If characterisation data are not disclosed by the sponsor to the contracted test facility, this fact should also be explicitly mentioned in the final report.

Submission of data to regulatory authorities

The ultimate responsibility for the scientific validity of a study lies with the Study Director, and not with the sponsor, whose responsibility is to make the decision, based on the outcome of the studies, whether or not to submit a chemical for registration to a regulatory authority.

9. THE APPLICATION OF THE OECD PRINCIPLES OF GLP TO *IN VITRO* STUDIES*

History

As efforts to decrease the use of animals in safety testing are intensifying, *in vitro* methods are gaining a more prominent role as alternatives or supplements to *in vivo* safety testing. Anticipated developments in the fields of toxicogenomics, toxicoproteomics, toxicometabonomics and in various high through-put screening techniques are expected to enhance the importance of *in vitro* methodologies for safety testing, beyond their traditional use as test systems in the area of genetic toxicity testing. The OECD Working Group on Good Laboratory Practice considered it therefore worthwhile to develop further guidance specifically of relevance to the application and interpretation of the OECD Principles of GLP to *in vitro* studies.

The Working Group established a Task Force under the leadership of Switzerland, which met in Bern on 12 to 13 February 2004. The Task Force comprised members of the Working Group or experts in *in vitro* testing nominated by them, representing Belgium, France, Germany, Japan, the Netherlands, Switzerland, the United States and the European Commission.

The draft Advisory Document developed by the Task Force was examined by the Working Group at its 18th Meeting in May 2004, where it was amended and endorsed. The Chemicals Committee and the Working Party on Chemicals, Pesticides and Biotechnology at its 37th Joint Meeting in turn endorsed the document in 2004. It is declassified under the authority of the Secretary-General.

Introduction

Studies involving *in vitro* test systems have long been used to obtain data on the safety of chemicals with respect to human health and the environment. National legislation usually requires that these studies be conducted in accordance with Good Laboratory Practice (GLP) requirements. Traditionally, *in vitro* methods have been mainly used in the area of genetic toxicity testing, where the hazard assessment is based to a large extent on data derived from studies using *in vitro* test systems. As efforts to decrease the use of animals in safety testing are intensifying, *in vitro* methods are gaining a more prominent role as alternatives or supplements to *in vivo* safety testing. Furthermore, developments in the area of toxicogenomics, toxicoproteomics, toxicometabonomics and various (*e.g.* micro-array) high through-put screening techniques will also enhance the importance of *in vitro* methodologies for safety testing.

The requirement that safety studies be planned, conducted, recorded, reported and archived in accordance with the OECD Principles of Good Laboratory Practice (hereafter the GLP Principles) does not differ for different study types. Therefore, the GLP Principles and

---

* No. 14 in the OECD Series on Principles of Good Laboratory Practice and Compliance Monitoring.

the associated Consensus Documents (Part II, Chapters 1-7 of this publication) describe requirements for and provide general guidance on the conduct of all non-clinical health and environmental safety studies, including *in vitro* studies. In order to facilitate the application and interpretation of the GLP Principles in relation to the specific *in vitro* testing situation, further clarification and guidance was considered useful.

Purpose of this document

The purpose of this document is to facilitate the proper application and interpretation of the GLP Principles for the organisation and management of *in vitro* studies, and to provide guidance for the appropriate application of the GLP Principles to *in vitro* studies, both for test facilities (management, QA, study director and personnel), and for national GLP compliance monitoring authorities.

This Advisory Document intends to provide such additional interpretation of the Principles and guidance for their application to *in vitro* studies carried out for regulatory purposes. It is organised in such a way as to provide easy reference to the GLP Principles by following the sequence of the different parts of these GLP Principles.

Scope

This document is specific to the application of the Principles of GLP to *in vitro* studies conducted in the framework of non-clinical safety testing of test items contained in pharmaceutical products, pesticide products, cosmetic products, veterinary drugs as well as food additives, feed additives, and industrial chemicals. These test items are frequently synthetic chemicals, but may be of natural or biological origin and, in some circumstances, may be living organisms. The purpose of testing these test items is to obtain data on their properties and/or their safety with respect to human health and/or the environment.

Unless specifically exempted by national legislation, the Principles of Good Laboratory Practice apply to all non-clinical health and environmental safety studies required by regulations for the purpose of registering or licensing pharmaceuticals, pesticides, food and feed additives, cosmetic products, veterinary drug products and similar products, and for the regulation of industrial chemicals.

Definitions

a) In vitro studies

In vitro studies are studies which do not use multicellular whole organisms, but rather microorganisms or material isolated from whole organisms, or simulations thereof as test systems.

Many *in vitro* studies will qualify as short-term studies under the definition provided by the GLP Principles. For these studies, the OECD Consensus Document on the Application of the GLP Principles to Short-Term Studies should be consulted and used as appropriate, in order to allow for the application of the provisions facilitating the work of Study Director and QA.

b) Reference item

Test guidelines for *in vitro* studies mandate in many cases the use of appropriate positive, negative and/or vehicle control items which may not serve, however, as the GLP definition of "reference items" implies, to grade the response of the test system to the test item, but rather to control the proper performance of the test system. Since the purpose of these positive, negative and/or vehicle control items may be considered as analogous to the purpose of a reference item, the definition of the latter may be regarded as covering the terms "positive, negative, and/or vehicle control items" as well. The extent to which they should be analytically characterized may, however, be different from the requirements of reference items.

Responsibilities

a) Test facility management

Most of the responsibilities of test facility management are of a general nature and are equally applicable to *in vivo* and *in vitro* studies, such as the requirements that test facility management has to ensure the availability of qualified personnel, and of appropriate facilities and equipment for the timely and proper conduct of the study. However, test facility management should be aware that *in vitro* testing may influence the execution of some of their responsibilities. For example, test facility management must ensure that personnel clearly understand the functions they are to perform. For *in vitro* studies, this may entail ensuring that specific training is provided in aseptic procedures and in the handling of biohazardous materials. *In vitro* testing may also necessitate the availability of specialized areas and the implementation of procedures to avoid contamination of test systems. Another example is provided by the requirement that test facility management should ensure that test facility supplies meet requirements appropriate to their use in a study. Certain *in vitro* studies may necessitate the use of proprietary materials or test kits. Although the OECD Consensus Document on Compliance of Laboratory Suppliers with GLP Principles states that materials to be used in a GLP compliant study should be produced and tested for suitability using an adequate quality system, thus placing the primary responsibility for their suitability on the manufacturer or supplier, it is the responsibility of the test facility management to confirm that these conditions are adequately fulfilled through assessment of the suppliers practices, procedures and policies.

b) Study Director

The general responsibilities of the Study Director are independent of the type of study and the responsibilities listed in the Principles apply to *in vitro* studies as well. The study director continues to be the single point of study control and has the responsibility for the overall conduct and reporting of the study.

In *in vitro* studies, the Study Director should pay particular attention to documenting the justification and characterisation of the test system, an activity which may be more difficult to accomplish for *in vitro* studies. See the section on Test Systems, below, regarding the documentation required to justify and characterise the test system. In *in vivo* studies, these activities have been rather straightforward. For example, the use of a particular species may be justified by documenting the characteristics of that species that make it an appropriate model for assessing the effect of interest. Characterisation of a particular animal may be accomplished by simply documenting the animal species, strain, substrain, source of supply, number, body weight range, sex, and age.

These required activities may be more difficult to accomplish for *in vitro* studies.

Justification of the test system may require that the Study Director document that the test method has been validated or is structurally, functionally, and/or mechanistically similar to a validated reference test method. Prior to the use of new test methods that are structurally, functionally and/or mechanistically similar to a validated reference test method, the Study Director should therefore provide documented evidence that the new test method has comparable performance when evaluated with appropriate reference items.

Characteristics of *in vitro* systems may also be difficult to document. Although the Study Director may be able, with the assistance of the supplier, to document some characteristics of the test system (*e.g.* cell line, age/passage, origin), he/she should also characterise the test system by documenting that the test system provides the required performance when evaluated with appropriate reference items, including positive, negative, and untreated and/or vehicle controls, where necessary. A special case may be seen in the use of proprietary materials or test kits in the conduct of *in vitro* studies. While the performance of such materials or test kits should be assured by the supplier, producer or patent holder, and while the test facility management is responsible for ensuring that the supplier meets the quality criteria as mentioned above, *e.g.* by reviewing vendor practices, procedures and policies, it is the responsibility of the Study Director to ensure that the performance of these materials or kits indeed meets the requirements of the study, and to ensure that test kits have been adequately validated and are suitable for their intended purpose. Since the quality and reliability of study results will be influenced directly by the quality and performance of these materials or test kits, it is especially important that the completeness and acceptability of the quality control documentation provided by the supplier should be thoroughly examined and critically evaluated by the Study Director. At a minimum, the Study Director should be able to judge the appropriateness of the quality system used by the manufacturer, and have available all documentation needed to assess the fitness for use of the test system, *e.g.* results of performance studies.

c) Study personnel

Personnel should meticulously observe, where applicable, the requirements for aseptic conditions and follow the respective procedures in the conduct of *in vitro* studies to avoid pathogen contamination of the test system. Similarly, personnel should employ adequate practices (see "Sources for further information", ref. 1) to avoid cross-contamination between test systems and to ensure the integrity of the study. Study personnel should be aware of, and strictly adhere to, the requirements to isolate test systems and studies involving biohazardous materials. Appropriate precautions to minimise risks originating from the use of hazardous chemicals should be applied during *in vitro* studies as well.

Quality Assurance

In general, Quality Assurance (QA) activities will not be greatly different between *in vitro* and *in vivo* studies. *In vitro* studies may qualify in certain cases for treatment under the conditions of short-term studies; in these cases, the OECD Consensus Document on the Application of the GLP Principles to Short-Term Studies will be applicable. Thus, such studies may be inspected, if applicable and permitted by national regulations, by QA on a process-based inspection programme. Since the GLP Principles require QA to inspect especially the critical phases of a study, it is important that, in the case of *in vitro* studies, QA is well aware of what constitutes critical phases (and critical aspects) of such studies. Corresponding guidance for QA inspections should be developed in co-operation with

Study Directors, Principal Investigators and study personnel in the relevant areas. Since the QA programme should, wherever indicated, explicitly cover specific aspects of *in vitro* testing, education and training of QA personnel should also be explicitly directed towards the ability to recognise potential problems in specific areas of *in vitro* testing.

Specific areas to be inspected may include, but not be limited to, the procedures and measures for:

- monitoring of batches of components of cell and tissue culture media that are critical to the performance of the test system (*e.g.* foetal calf serum, etc.) and other materials with respect to their influence on test system performance;
- assessing and ensuring functional and/or morphological status (and integrity) of cells, tissues and other indicator materials;
- monitoring for potential contamination by foreign cells, mycoplasma and other pathogens, or other adventitious agents, as appropriate;
- cleaning and decontamination of facilities and equipment and minimising sources of contamination of test items and test systems;
- ensuring that specialised equipment is properly used and maintained;
- ensuring proper cryopreservation and reconstitution of cells and tissues;
- ensuring proper conditions for retrieval of materials from frozen storage;
- ensuring sterility of materials and supplies used for cell and tissue cultures;
- maintaining adequate separation between different studies and test systems.

Facilities

a) General

The GLP Principles mandate that test facilities should be suitable to meet the requirements of the studies performed therein, and that an adequate degree of separation should be maintained between different activities to ensure the proper and undisturbed conduct of each study. Due to the fact that *in vitro* studies generally occupy only limited workspace and do not normally require dedicated facilities that exclude the performance of other studies, measures should be taken to ensure the appropriate separation of *in vitro* studies co-existing in the same physical environment.

b) Test system facilities

The GLP Principles require that a sufficient number of rooms or areas should be available to ensure the isolation of test systems, and that such areas should be suitable to ensure that the probability of contamination of test systems is minimised. The term "areas", however, is not specifically defined and its interpretation is therefore adaptable to various *in vitro* situations. The important aspect here is that the integrity of each test system and study should not be jeopardised by the possibility of potential contamination or cross-contamination or mix-up.

In this way, it may be possible to incubate cells or tissues belonging to different studies within the same incubator, provided that an adequate degree of separation exists (*e.g.* appropriate identifiers, labelling or separate placement to distinguish between studies, etc.), and that no test item is sufficiently volatile so as to contaminate other studies that are co-incubated.

Separation of critical study phases may be possible not only on a spatial, but also on a temporal basis. Manipulation of cell and tissue cultures, *e.g.* subcultivation procedures, addition of test item, etc., is normally performed in vertical laminar flow cabinets to assure sterility and to protect the test system as well as study personnel and the environment. Under these circumstances, adequate separation to prevent cross-contamination between different studies will be achieved by sequential manipulation of the test systems used in the individual studies, with careful cleaning and decontamination/sterilisation of the working surfaces of the cabinet and of related laboratory equipment performed between the different activities, as necessary.

Another important aspect is the availability of devoted rooms or areas with special equipment for the long-term storage of test systems. The equipment, including storage containers, should provide adequate conditions for maintenance of long-term integrity of test systems.

c) Facilities for handling test and reference items

While the requirements of the GLP Principles for handling test and reference items apply equally to *in vitro* tests as far as the prevention of cross-contamination by test and reference items is concerned, another aspect needs to be taken into account: Since sterility is an important consideration in *in vitro* studies, it should be ensured that rooms or areas used for preparation and mixing of test and reference items with vehicles be equipped so as to allow working under aseptic conditions, and thus protecting the test system and the study by minimising the probability of their contamination by test and reference item preparations.

Apparatus, material, and reagents

While the commonly observed, routine requirements for apparatus used in a GLP compliant environment apply equally to apparatus used for *in vitro* studies, there are some specific points and issues of particular importance. As an example, it may be of importance for the integrity and reliability of some *in vitro* studies to ensure that the proper conditions of certain equipment, like microbalances, micropipettes, laminar air flow cabinets or incubators are regularly maintained, and monitored and calibrated where applicable. For specific equipment, critical parameters should be identified requiring continuous monitoring or the setting of limit values together with installation of alarms.

The requirements in the GLP Principles for reagents with respect to labelling and expiry dates apply equally to those used for *in vitro* studies.

Test systems

In vitro test systems are mainly biological systems, although some of the more recent developments in alternatives to conventional *in vivo* testing (*e.g.* gene arrays for toxicogenomics) may also exhibit some attributes of physical-chemical test systems, and still others, *e.g.* toxicometabonomics, may mainly rely on analytical methodology. Test kits, including proprietary test kits, should also be considered as test systems.

a) Conditions for test systems

As for any other biological test systems, adequate conditions should be defined, maintained and monitored to ensure the quality and integrity of the test system during storage and within the study itself. This includes the documented definition, maintenance and monitoring of the viability and responsiveness of the test system, including recording of cell passage number and population doubling times. Records should also be kept for environmental conditions (*e.g.* liquid nitrogen level in a liquid nitrogen cryostorage system, temperature, humidity and CO_2 concentration in incubators, etc.) as well as for any manipulation of the test system required for the maintenance of its quality and integrity (*e.g.* treatment with antibiotics or antifungals, subcultivation, selective cultivation for reducing the frequency of spontaneous events). Since maintenance of the proper environmental conditions during the storage of test systems may influence data quality to a greater degree than for other biological systems, these records may be of special importance in the maintenance of data quality and reliability.

b) Newly received test systems

Documentation obtained from the supplier of *in vitro* test systems (*e.g.* origin, age/passage number, cell doubling time and other relevant characteristics that help identify the test system) should be reviewed and retained in the study records. Predefined criteria should be used to assess the viability, suitability (*e.g.* functional and/or morphological status of cells and tissues, testing for known or suspected microbial or viral contaminants) and responsiveness of the test system. Results of such evaluations should be documented and retained in the study records. If no such assessment is possible, as, *e.g.*, with primary cell cultures or "reconstituted organs", a mechanism should exist between the supplier and the user to ascertain and document the suitability of the test system. Monitoring and recording performance against negative and positive control items may constitute sufficient proof for the responsiveness of a given test system. Any problems with the test system that may affect the quality, validity and reliability of the study should be documented and discussed in the final report. Problems with vendor-supplied test systems should be brought to the attention of the vendor and corrective measures sought.

c) Test system records

The GLP Principles require that records be maintained of source, date of arrival and arrival condition of test systems; for cells and tissues, these records should include not only the immediate source (*e.g.* commercial supplier), but also the original source from where the cells or tissues have been derived (*e.g.* primary cells or tissues with donor characteristics; established cell lines from recognised sources, etc.). Other information to be maintained should include, but not be limited to, the method by which cells or tissues were originally obtained (*e.g.* derived from tissue explants, biopsies of normal or cancer tissues, gene transfer by plasmid transfection or virus transduction, etc.), chronology of custody, passage number of cell lines, culture conditions and subcultivation intervals, freezing/thawing conditions, etc. For transgenic test systems, it is necessary, in addition, to ascertain the nature of the transgene and to monitor maintenance of expression with appropriate controls.

Special attention should be paid to the proper labelling of test systems during storage and use, which includes measures to ensure the durability of labelling. Especially where the size of containers and the conditions of storage (*e.g.* cryovials in liquid nitrogen, multiple test systems stored in one container) may be critical factors for labelling, measures should be in place to ensure the correct identification of test systems at all times.

The requirements in the OECD Principles of GLP for test items and reagents with respect to labelling and expiry dates apply equally to test kits used as *in vitro* test systems. Test kits, whether used as test systems or in any other way, *e.g.* for analytical purposes, should have an expiry date. Extending this expiry date can be only acceptable on the basis of documented evaluation (or analysis). For test kits used as test systems, such documented evaluation may, *e.g.* consist of the historical record of observed responses obtained with the respective batch of the test kit to positive, negative and/or vehicle control items, and proof that, even after the expiry date, the response did not deviate from the historical control values. A documented decision of the Study Director as to the extension of the expiry date should provide evidence for this evaluation process.

In order to avoid possible confusion, the nomenclature for the test systems should be clearly defined, and test system labels as well as all records obtained from individual studies should bear the formally accepted designation of the test system.

Test and reference items (including negative and positive control items)

In general, there are no specific requirements for receipt, handling, sampling, storage and characterisation for test and reference items that are used in studies utilising *in vitro* test systems besides those listed in the GLP Principles. Aseptic conditions may, however, be required in their handling to avoid microbial contamination of test systems.

For negative, vehicle and positive control items, it may or may not be necessary to determine concentration and homogeneity, since it may be sufficient to provide evidence for the correct, expected response of the test system to them.

The expiry date of such control items may also be extended by documented evaluation or analysis. Such evaluation may consist of documented evidence that the response of the respective test systems to these positive, negative and/or vehicle control items does not deviate from the historical control values recorded in the test facility, which should furthermore be comparable to published reference values.

Standard Operating Procedures (SOPs)

In addition to the examples cited in the GLP Principles (see Section 7.4.1-7.4.5), there are activities and processes specific to *in vitro* testing that should be described in Standard Operating Procedures. Such SOPs should therefore be additionally available for, but not be limited to, the following illustrative examples for test facility activities related to *in vitro* testing.

a) Facilities

Environmental monitoring with respect to pathogens in the air and on surfaces, cleaning and disinfection, actions to take in the case of infection or contamination in the test facility or area.

b) Apparatus

Use, maintenance, performance monitoring, cleaning, and decontamination of cell and tissue culture equipment and instruments, such as laminar-flow cabinets and incubators; monitoring of liquid nitrogen levels in storage containers; calibration and monitoring of temperature, humidity and CO_2-levels in incubators.

c) Materials, reagents and solutions

Evaluation of suitability, extension of expiry dates, assessment and maintenance of sterility, screening for common pathogen contaminants; description of procedures for choice and use of vehicles; verification procedures for compatibility of vehicles with the test system.

d) Test systems

Conditions for storage and procedures for freezing and thawing of cells and tissues, testing for common pathogens; visual inspection for contaminations; verification procedures (*e.g.* use of acceptance criteria) for ensuring properties and responsiveness on arrival and during use, whether immediately after arrival or following storage; morphological evaluation, control of phenotype or karyotype stability, control of transgene stability; mode of culture initiation, culture conditions with subcultivation intervals; handling of biohazardous materials and test systems, procedures for disposal of test systems.

e) Performance of the study

Aseptic techniques, acceptance criteria for study validity, criteria for assay repetitions.

f) Quality assurance

Definition of critical phases, inspection frequencies.

Performance of the study and reporting of study results

The GLP requirements for the performance of *in vitro* studies are identical to those provided for the more conventional safety studies. In many cases, the OECD Consensus Document on the Application of the GLP Principles to Short-Term Studies may be consulted in combination with the OECD GLP Principles in order that *in vitro* studies may be conducted in a GLP compliant way.

There are a number of issues specific to *in vitro* testing that should be addressed in the study plan as well as in the final study report. These issues, however, are mainly of a scientific, technical nature, such as the (scientific) requirement that any internal controls (appropriate positive, negative, and untreated and/or vehicle controls), carried out in order to control bias and to evaluate the performance of the test system, should be conducted concurrently with the test item in all *in vitro* studies. More specific guidance as to what topics should be addressed in the study plan and the final report will be found in the respective OECD test guidelines or other appropriate references.

Storage and retention of records and materials

The general retention requirements of the GLP Principles apply to *in vitro* studies as well. Additionally, it should be considered to retain samples of long-term preservable test systems, especially test systems of limited availability (*e.g.* special subclones of cell lines, transgenic cells, etc.), in order to enable confirmation of test system identity, and/or for study reconstructability.

Retention of samples of test item should be considered also for such *in vitro* studies which can be categorised as short-term studies, especially in cases where *in vitro* studies constitute the bulk of safety studies.

Records of historical positive, negative, and untreated and/or vehicle control results used to establish the acceptable response range of the test system should also be retained.

Glossary of terms

Within the context of this document the following definitions are used:

Aseptic conditions: Conditions provided for, and existing in, the working environment under which the potential for microbial and/or viral contamination is minimised.

Cell lines: Cells that have undergone a genetic change to immortalisation and that, in consequence, are able to multiply for extended periods *in vitro*, and can be expanded and cryopreserved as cell bank deposits. A continuous cell line is generally more homogeneous, more stable, and thus more reproducible than a heterogeneous population of primary cells.

Control, negative: Separate part of a test system treated with an item, for which it is known that the test system should not respond; the negative control provides evidence that the test system is not responsive under the actual conditions of the assay.

Control, positive: Separate part of the test system treated with an item, the response to which is known for the test system; the positive control provides evidence that the test system is responsive under the actual conditions of the assay.

Control, untreated: Separate untreated part of a test system that is kept under the original culture conditions; the untreated control provides baseline data of the test system under the conditions of the assay.

Control, vehicle: Separate part of a test system to which the vehicle for the test item is added; the vehicle control provides evidence for a lack of influence of the chosen vehicle on the test system under the actual conditions of the assay.

Critical phases: Individual, defined procedures or activities within a study, on the correct execution of which the study quality, validity and reliability is critically dependent.

Cross-contamination: Contamination of a test item by another test item or of a test system by another test item or by another test system that is introduced inadvertently and taints the test item or impairs the test system.

Cryopreservation: Storage of cells and tissues by keeping them frozen under conditions where their viability is preserved.

Cryovial: Special vial used for cryopreservation. A cryovial has to satisfy special conditions such as tightness of closure even at extremely low temperatures and extreme temperature changes encountered during freezing and thawing.

Ex vivo: Cells, tissues, or organs removed for further analysis from intact animals.

Gene transfection: The introduction of foreign, supplemental DNA (single or multiple genes) into a host cell.

High through-put screening: The use of miniaturised, robotics-based technology to screen large compound libraries against an isolated target gene, protein, cell, tissue, etc. to select compounds on the basis of specific activities for further development.

Micro-arrays: Sets of miniaturised chemical reaction areas arranged in an orderly fashion and spotted onto a solid matrix such as a microscope slide. A DNA microarray provides a medium for matching known and unknown DNA samples based on base-pairing rules and allows for the automation of the process of identifying unknown DNA samples for use in probing a biological sample to determine gene expression, marker pattern or nucleotide sequence of DNA/RNA.

Primary cells: Cells that are freshly isolated from animal or plant sources. Freshly isolated primary cells may rapidly dedifferentiate in culture, and they often have a limited lifespan. Primary cultures isolated from animals or humans may represent heterogeneous populations with respect, for example, to differences in cell types and states of differentiation depending on purification techniques used. Each isolate will be unique and impossible to reproduce exactly. Primary cell cultures commonly require complex nutrient media, supplemented with serum and other components. Consequently, primary cell culture systems are extremely difficult to standardise.

Proprietary material: Material protected by (patent, copyright, or trademark) laws from illicit use.

Test kit: Ready-to-use compilation of all components necessary for the performance of an assay, test or study.

Tissues: Multicellular aggregates of differentiated cells with specific function as constituents of organisms.

Toxicogenomics: The study of how genomes respond to environmental stressors or toxicants. The goal of toxicogenomics is to find correlations between toxic responses to toxicants and changes in the genetic profiles of the objects exposed to such toxicants. Toxicogenomics combines the emerging technologies of genomics and bioinformatics to identify and characterise mechanisms of action of known and suspected toxicants. Currently, the premier toxicogenomic tools are the DNA microarray and the DNA chip, which are used for the simultaneous monitoring of expression levels of hundreds to thousands of genes.

Toxicometabonomics: The quantitative measurement of the time-related multiparametric metabolic response of living systems to pathophysiological stimuli or genetic modification by the systematic exploration of biofluid composition using NMR/pattern recognition technology in order to associate target organ toxicity with NMR spectral patterns and identify novel surrogate markers of toxicity.

Toxicoproteomics: The study of how the global protein expression in a cell or tissue responds to environmental stressors or toxicants. The goal of toxicoproteomics is to find correlations between toxic responses to toxicants and changes in the complete complements of proteins profiles of the objects exposed to such toxicants.

Transgenic cells: Cells transfected with one or more foreign gene(s) which consequently express characteristics and functions that are normally not present, or at low expression levels only, in the parental cell.

Sources for further tnformation on *in vitro* testing

Web pages of:

1. Good Cell Culture Practices
   http://ecvam.jrc.it/publication/index5007.html
2. MIAME Guideline
   www.mged.org/Workgroups/MIAME/miame.html
3. ECVAM
   http://ecvam.jrc.it/index.htm
4. ICCVAM
   http://iccvam.niehs.nih.gov/

PART III

Guidance and Advisory Documents for Good Laboratory Practice Compliance Monitoring Authorities

The first two Guidance Documents for GLP Compliance Monitoring Authorities reproduced here are integral parts of the 1989 Council Decision-Recommendation on Compliance with Good Laboratory Practice [C(89)87(Final), Annexes I and II]. They were adopted by the OECD Council to assist member countries to implement the recommendation out in Part II, Paragraph 1 of that Council Act.

The other two documents in this section were developed by the Working Group on GLP. They were endorsed by the Joint Meeting of the Chemicals Committee and Working Party on Chemicals, Pesticides and Biotechnology and declassified under the authority of the Secretary-General.

PART III

Chapter 3

Guidance for GLP Monitoring Authorities

1. GUIDES FOR COMPLIANCE MONITORING PROCEDURES FOR GOOD LABORATORY PRACTICE*

History

The 1981 Council Decision on Mutual Acceptance of Data [C(81)30(Final)], of which the OECD Principles of Good Laboratory Practice are an integral part, includes an instruction for OECD to undertake activities "to facilitate internationally-harmonised approaches to assuring compliance" with the GLP Principles. Consequently, in order to promote the implementation of comparable compliance monitoring procedures, and international acceptance, among member countries, the Council adopted in 1983 the Recommendation concerning the Mutual Recognition of Compliance with Good Laboratory Practice [C(83)95(Final)], which set out basic characteristics of the procedures for monitoring compliance.

A Working Group on Mutual Recognition of Compliance with GLP was established in 1985, under the chairmanship of Professor V. Silano (Italy), to facilitate the practical implementation of the Council acts on GLP, develop common approaches to the technical and administrative problems related to GLP compliance and its monitoring, and develop arrangements for the mutual recognition of compliance monitoring procedures. The following countries and organisations participated in the Working Group: Australia, Belgium, Canada, Denmark, the Federal Republic of Germany, Finland, France, Italy, Japan, Norway, the Netherlands, Portugal, Spain, Sweden, Switzerland, the United Kingdom, the United States, the Commission of the European Communities, the International Organisation for Standardisation, the Pharmaceuticals Inspections Convention, and the World Health Organization.

The Working Group developed, *inter alia*, Guides for Compliance Monitoring Procedures for Good Laboratory Practice, which concern the requisites of administration, personnel and GLP compliance monitoring programmes. These were first published in 1988 in the Final Report of the Working Group on Mutual Recognition of Compliance with Good Laboratory Practice. A slightly abridged version was annexed to the 1989 Council Decision-Recommendation on Compliance with Principles of Good Laboratory Practice [C(89)87(Final)], which superseded and replaced the 1983 Council Act.

In adopting that Decision-Recommendation, the Council in Part III.1 instructed the Environment Committee and the Management Committee of the Special Programme on the Control of Chemicals to ensure that the "Guides for Compliance Monitoring Procedures for Good Laboratory Practice" and the "Guidance for the Conduct of Laboratory Inspections and Study Audits" set out in Annexes I and II thereto were updated and expanded, as necessary, in light of developments and experience of member countries and relevant work in other international organisations.

---

* No. 2 in the OECD Series on Principles of Good Laboratory Practice and Compliance Monitoring.

The OECD Panel on Good Laboratory Practice (later named Working Group on GLP) developed proposals for amendments to these Annexes, as well as to Annex III which provides "Guidance for the Exchange of Information concerning National Programmes for Monitoring of Compliance with the Principles of Good Laboratory Practice" and which was amended essentially to include an appendix on "Guidance for Good Laboratory Practice Monitoring Authorities for the Preparation of Annual Overviews of Test Facilites Inspected". These revised Annexes were approved by the Council in a Decision Amending the Annexes to the Council Decision – Recommendation on Compliance with Principles of Good Laboratory Practice on 9 March 1995 [C(95)8(Final)].

The "Revised Guides for Compliance Monitoring Procedures for Good Laboratory Practice" are contained in the revision of Annex I to the "Council Decision – Recommendation on Compliance with Principles of Good Laboratory Practice [C(89)87(Final)]", [C(95)8(Final) adopted by the Council on 9 March 1995]. [For the text of C(89)87(Final), see the Annex, Section 2, of this publication.]

Introduction

To facilitate the mutual acceptance of test data generated for submission to Regulatory Authorities of OECD member countries, harmonisation of the procedures adopted to monitor good laboratory practice compliance, as well as comparability of their quality and rigour, are essential. The aim of this document is to provide detailed practical guidance to OECD member countries on the structure, mechanisms and procedures they should adopt when establishing national Good Laboratory Practice compliance monitoring programmes so that these programmes may be internationally acceptable.

It is recognised that member countries will adopt GLP Principles and establish compliance monitoring procedures according to national legal and administrative practices, and according to priorities they give to, *e.g.*, the scope of initial and subsequent coverage concerning categories of chemicals and types of testing. Since member countries may establish more than one Good Laboratory Practice Monitoring Authority due to their legal framework for chemicals control, more than one Good Laboratory Practice Compliance Programme may be established. The guidance set forth in the following paragraphs concerns each of these Authorities and Compliance Programmes, as appropriate.

Definitions of terms

The definitions of terms in the OECD Principles of Good Laboratory Practice [Annex 2 to Council Decision C(81)30(Final)] are applicable to this document. In addition, the following definitions apply:

GLP Principles: Principles of good laboratory practice that are consistent with the OECD Principles of Good Laboratory Practice as set out in Annex 2 of Council Decision C(81)30(Final).

GLP Compliance Monitoring: The periodic inspection of test facilities and/or auditing of studies for the purpose of verifying adherence to GLP Principles.

(National) GLP Compliance Programme: The particular scheme established by a member country to monitor good laboratory practice compliance by test facilities within its territories, by means of inspections and Study Audits.

(National) GLP Monitoring Authority: A body established within a member country with responsibility for monitoring the good laboratory practice compliance of test facilities within its territories and for discharging other such functions related to good laboratory

practice as may be nationally determined. It is understood that more than one such body may be established in a member country.

Test Facility Inspection: An on-site examination of the test facility's procedures and practices to assess the degree of compliance with GLP Principles. During inspections, the management structures and operational procedures of the test facility are examined, key technical personnel are interviewed, and the quality and integrity of data generated by the facility are assessed and reported.

Study Audit: A comparison of raw data and associated records with the interim or final report in order to determine whether the raw data have been accurately reported, to determine whether testing was carried out in accordance with the study plan and Standard Operating Procedures, to obtain additional information not provided in the report, and to establish whether practices were employed in the development of data that would impair their validity.

Inspector: A person who performs the Test Facility Inspections and Study Audits on behalf of the (National) GLP Monitoring Authority.

GLP Compliance Status: The level of adherence of a test facility to the GLP Principles as assessed by the (National) GLP Monitoring Authority.

Regulatory Authority: A national body with legal responsibility for aspects of the control of chemicals.

Components of Good Laboratory Practice compliance monitoring procedures

Administration

A (National) GLP Compliance Programme should be the responsibility of a properly constituted, legally identifiable body adequately staffed and working within a defined administrative framework.

Member countries should:

- ensure that the (National) GLP Monitoring Authority is directly responsible for an adequate "team" of inspectors having the necessary technical/scientific expertise or is ultimately responsible for such a "team";
- publish documents relating to the adoption of GLP Principles within their territories;
- publish documents providing details of the (National) GLP Compliance Programme, including information on the legal or administrative framework within which the programme operates and references to published acts, normative documents (*e.g.* regulations, codes of practice), inspection manuals, guidance notes, periodicity of inspections and/or criteria for inspection schedules, etc.;
- maintain records of test facilities inspected (and their GLP Compliance Status) and of studies audited for both national and international purposes.

Confidentiality

(National) GLP Monitoring Authorities will have access to commercially valuable information and, on occasion, may even need to remove commercially sensitive documents from a test facility or refer to them in detail in their reports.

Member countries should:

- make provision for the maintenance of confidentiality, not only by Inspectors but also by any other persons who gain access to confidential information as a result of GLP Compliance Monitoring activities;
- ensure that, unless all commercially sensitive and confidential information has been excised, reports of Test Facility Inspections and Study Audits are made available only to Regulatory Authorities and, where appropriate, to the test facilities inspected or concerned with Study Audits and/or to study sponsors.

Personnel and training

(National) GLP Monitoring Authorities should:

- *ensure that an adequate number of Inspectors is available*

  The number of Inspectors required will depend upon:

  i) the number of test facilities involved in the (National) GLP Compliance Programme;
  ii) the frequency with which the GLP Compliance Status of the test facilities is to be assessed;
  iii) the number and complexity of the studies undertaken by those test facilities;
  iv) the number of special inspections or audits requested by Regulatory Authorities.

- *ensure that Inspectors are adequately qualified and trained*

  Inspectors should have qualifications and practical experience in the range of scientific disciplines relevant to the testing of chemicals. (National) GLP Monitoring Authorities should:

  i) ensure that arrangements are made for the appropriate training of GLP Inspectors, having regard to their individual qualifications and experience;
  ii) encourage consultations, including joint training activities where necessary, with the staff of (National) GLP Monitoring Authorities in other member countries in order to promote international harmonisation in the interpretation and application of GLP Principles, and in the monitoring of compliance with such Principles.

- *ensure that inspectorate personnel, including experts under contract, have no financial or other interests in the test facilities inspected, the studies audited or the firms sponsoring such studies;*
- *provide Inspectors with a suitable means of identification (e.g. an identity card).*

  Inspectors may be:

- on the permanent staff of the (National) GLP Monitoring Authority;
- on the permanent staff of a body separate from the (National) GLP Monitoring Authority; or
- employed on contract, or in another way, by the (National) GLP Monitoring Authority to perform Test Facility Inspections or Study Audits.

  In the latter two cases, the (National) GLP Monitoring Authority should have ultimate responsibility for determining the GLP Compliance Status of test facilities and the quality/acceptability of a Study Audit, and for taking any action based on the results of Test Facility Inspections or Study Audits which may be necessary.

(National) GLP Compliance Programmes

GLP Compliance Monitoring is intended to ascertain whether test facilities have implemented GLP Principles for the conduct of studies and are capable of assuring that the resulting data are of adequate quality. As indicated above, member countries should publish the details of their (National) GLP Compliance Programmes. Such information should, *inter alia*:

- *define the scope and extent of the Programme*

    A (National) GLP Compliance Programme may cover only a limited range of chemicals, *e.g.* industrial chemicals, pesticides, pharmaceuticals, etc., or may include all chemicals. The scope of the monitoring for compliance should be defined, both with respect to the categories of chemicals and to the types of tests subject to it, *e.g.* physical, chemical, toxicological and/or ecotoxicological.

- *provide an indication as to the mechanism whereby test facilities enter the Programme*

    The application of GLP Principles to health and environmental safety data generated for regulatory purposes may be mandatory. A mechanism should be available whereby test facilities may have their compliance with GLP Principles monitored by the appropriate (National) GLP Monitoring Authority.

- *provide information on categories of Test Facility Inspections/Study Audits*

    A (National) GLP Compliance Programme should include:

    i) provision for Test Facility Inspections. These inspections include both a general Test Facility Inspection and a Study Audit of one or more on-going or completed studies;

    ii) provision for special Test Facility Inspections/Study Audits at the request of a Regulatory Authority – *e.g.* prompted by a query arising from the submission of data to a Regulatory Authority.

- *define the powers of Inspectors for entry into test facilities and their access to data held by test facilities (including specimens, SOP's, other documentation, etc.)*

    While Inspectors will not normally wish to enter test facilities against the will of the facility's management, circumstances may arise where test facility entry and access to data are essential to protect public health or the environment. The powers available to the (National) GLP Monitoring Authority in such cases should be defined.

- *describe the Test Facility Inspection and Study Audit procedures for verification of GLP compliance*

    The documentation should indicate the procedures which will be used to examine both the organisational processes and the conditions under which studies are planned, performed, monitored and recorded. Guidance for such procedures is available in Guidance for the Conduct of Test Facility Inspections and Study Audits (No. 3 in the OECD series on Principles of GLP and Compliance Monitoring).

- *describe actions that may be taken as follow-up to Test Facility Inspections and Study Audits.*

Follow-up to Test Facility Inspections and Study Audits

When a Test Facility Inspection or Study Audit has been completed, the Inspector should prepare a written report of the findings.

Member countries should take action where deviations from GLP Principles are found during or after a Test Facility Inspection or Study Audit. The appropriate actions should be described in documents from the (National) GLP Monitoring Authority.

If a Test Facility Inspection or Study Audit reveals only minor deviations from GLP Principles, the facility should be required to correct such minor deviations. The Inspector may need, at an appropriate time, to return to the facility to verify that corrections have been introduced.

Where no or where only minor deviations have been found, the (National) GLP Monitoring Authority may:

- issue a statement that the test facility has been inspected and found to be operating in compliance with GLP Principles. The date of the inspections and, if appropriate, the categories of test inspected in the test facility at that time should be included. Such statements may be used to provide information to (National) GLP Monitoring Authorities in other member countries;

  and/or

- provide the Regulatory Authority which requested a Study Audit with a detailed report of the findings.

Where serious deviations are found, the action taken by (National) GLP Monitoring Authorities will depend upon the particular circumstances of each case and the legal or administrative provisions under which GLP Compliance Monitoring has been established within their countries. Actions which may be taken include, but are not limited to, the following:

- issuance of a statement, giving details of the inadequacies or faults found which might affect the validity of studies conducted in the test facility;
- issuance of a recommendation to a Regulatory Authority that a study be rejected;
- suspension of Test Facility Inspections or Study Audits of a test facility and, for example and where administratively possible, removal of the test facility from the (National) GLP Compliance Programme or from any existing list or register of test facilities subject to GLP Test Facility Inspections;
- requiring that a statement detailing the deviations be attached to specific study reports;
- action through the courts, where warranted by circumstances and where legal/administrative procedures so permit.

Appeals procedures

Problems, or differences of opinion, between Inspectors and test facility management will normally be resolved during the course of a Test Facility Inspection or Study Audit. However, it may not always be possible for agreement to be reached. A procedure should exist whereby a test facility may make representations relating to the outcome of a Test Facility Inspection or Study Audit for GLP Compliance Monitoring and/or relating to the action the GLP Monitoring Authority proposes to take thereon.

2. GUIDANCE FOR THE CONDUCT OF TEST FACILITY INSPECTIONS AND STUDY AUDITS*

History

The 1981 Council Decision on Mutual Acceptance of Data [C(81)30(Final)], of which the OECD Principles of Good Laboratory Practice are an integral part, includes an instruction for OECD to undertake activities "to facilitate internationally-harmonised approaches to assuring compliance" with the GLP Principles. Consequently, in order to promote the implementation of comparable compliance monitoring procedures and international acceptance among member countries, the Council adopted in 1983 the Recommendation concerning the Mutual Recognition of Compliance with Good Laboratory Practice [C(83)95(Final)], which set out basic characteristics of the procedures for monitoring compliance.

A Working Group on Mutual Recognition of Compliance with GLP was established in 1985 under the chairmanship of Professor V. Silano (Italy) to facilitate the practical implementation of the Council acts on GLP, develop common approaches to the technical and administrative problems related to GLP compliance and its monitoring, and develop arrangements for the mutual recognition of compliance monitoring procedures. The following countries and organisations participated in the Working Group: Australia, Belgium, Canada, Denmark, the Federal Republic of Germany, Finland, France, Italy, Japan, Norway, the Netherlands, Portugal, Spain, Sweden, Switzerland, the United Kingdom, the United States, the Commission of the European Communities, the International Organisation for Standardisation, the Pharmaceutical Inspection Convention, and the World Health Organization.

The Working Group developed, *inter alia*, Guidance for the Conduct of Laboratory Inspections and Study Audits. The Guidance was based on a text developed by the Expert Group on GLP and presented as part of its Final Report in 1982 ("Good Laboratory Practice in the Testing of Chemicals", OECD, 1982, out of print). The current Guidance was first published in 1988 in the "Final Report of the Working Group on Mutual Recognition of Compliance with Good Laboratory Practice" ("OECD Environment Monograph No. 15", March 1988, out of print). A slightly abridged version was annexed to the 1989 Council Decision-Recommendation on Compliance with Principles of Good Laboratory Practice [C(89)87(Final)], which superseded and replaced the 1983 Council Act.

In adopting that Decision-Recommendation, the Council in Part III.1 instructed the Environment Committee and the Management Committee of the Special Programme on the Control of Chemicals to ensure that the "Guides for Compliance Monitoring Procedures for Good Laboratory Practice" and the "Guidance for the Conduct of Laboratory Inspections and Study Audits" set out in Annexes I and II thereto were updated and expanded, as

---

* No. 3 in the OECD Series on Principles of Good Laboratory Practice and Compliance Monitoring.

necessary, in light of developments and experience of member countries and relevant work in other international organisations.

The OECD Panel on Good Laboratory Practice developed proposals for amendments to these Annexes. These revised Annexes were approved by the Council in a Decision Amending the Annexes to the Council Decision-Recommendation on Compliance with Principles of Good Laboratory Practice on 9 March 1995 [C(95)8(Final)].

The "Revised Guidance for the Conduct of Laboratory Inspections and Study Audits" is contained in the revision of Annex II to the Council Decision-Recommendation on Compliance with Principles of Good Laboratory Practice [C(89)87(Final)], [C(95)8(Final) adopted by the Council on 9 March 1995]. [For the text of C(89)87(Final), see the Annex, Section 2, of this publication.]

Introduction

The purpose of this document is to provide guidance for the conduct of Test Facility Inspections and Study Audits which would be mutually acceptable to OECD member countries. It is principally concerned with Test Facility Inspections, an activity which occupies much of the time of GLP Inspectors. A Test Facility Inspection will usually include a Study Audit or "review" as a part of the inspection, but Study Audits will also have to be conducted from time to time at the request, for example, of a Regulatory Authority. General guidance for the conduct of Study Audits will be found at the end of this document.

Test Facility Inspections are conducted to determine the degree of conformity of test facilities and studies with GLP Principles and to determine the integrity of data to assure that resulting data are of adequate quality for assessment and decision-making by national Regulatory Authorities. They result in reports which describe the degree of adherence of a test facility to the GLP Principles. Test Facility Inspections should be conducted on a regular, routine basis to establish and maintain records of the GLP compliance status of test facilities.

Further clarification of many of the points in this document may be obtained by referring to the OECD Consensus Documents on GLP (on, *e.g.*, the role and responsibilities of the Study Director).

Definitions of terms

The definitions of terms in the OECD Principles of Good Laboratory Practice [Annex II to Council Decision C(81)30(Final)] and in the "Guides for Compliance Monitoring Procedures for Good Laboratory Practice" (Part III, chapter 1 of this publication) [Annex I to Council Decision-Recommendation C(89)87(Final)/revised in C(95)8(Final)] are applicable to this document.

Test Facility Inspections

Inspections for compliance with GLP Principles may take place in any test facility generating health or environmental safety data for regulatory purposes. Inspectors may be required to audit data relating to the physical, chemical, toxicological or ecotoxicological properties of a substance or preparation. In some cases, Inspectors may need assistance from experts in particular disciplines.

The wide diversity of facilities (in terms both of physical layout and management structure), together with the variety of types of studies encountered by Inspectors, means that the Inspectors must use their own judgment to assess the degree and extent of compliance with GLP Principles. Nevertheless, Inspectors should strive for a consistent approach in evaluating whether, in the case of a particular test facility or study, an adequate level of compliance with each GLP Principle has been achieved.

In the following sections, guidance is provided on the various aspects of the testing facility, including its personnel and procedures, which are likely to be examined by Inspectors. In each section, there is a statement of purpose, as well as an illustrative list of specific items which could be considered during the course of a Test Facility Inspection. These lists are not meant to be comprehensive and should not be taken as such.

Inspectors should not concern themselves with the scientific design of the study or the interpretation of the findings of studies with respect to risks for human health or the environment. These aspects are the responsibility of those Regulatory Authorities to which the data are submitted for regulatory purposes.

Test Facility Inspections and Study Audits inevitably disturb the normal work in a facility. Inspectors should therefore carry out their work in a carefully planned way and, so far as practicable, respect the wishes of the management of the test facility as to the timing of visits to certain sections of the facility.

Inspectors will, while conducting Test Facility Inspections and Study Audits, have access to confidential, commercially valuable information. It is essential that they ensure that such information is seen by authorised personnel only. Their responsibilities in this respect will have been established within their (National) GLP Compliance Monitoring Programme.

Inspection procedures

Pre-inspection

PURPOSE: To familiarise the Inspector with the facility which is about to be inspected in respect of management structure, physical layout of buildings and range of studies.

Prior to conducting a Test Facility Inspection or Study Audit, Inspectors should familiarise themselves with the facility which is to be visited. Any existing information on the facility should be reviewed. This may include previous inspection reports, the layout of the facility, organisation charts, study reports, protocols and curricula vitae (CVs) of personnel. Such documents would provide information on:

- the type, size and layout of the facility;
- the range of studies likely to be encountered during the inspection;
- the management structure of the facility.

Inspectors should note, in particular, any deficiencies from previous Test Facility Inspections. Where no previous Test Facility Inspections have been conducted, a pre-inspection visit can be made to obtain relevant information.

Test Facilities may be informed of the date and time of Inspector's arrival, the objective of their visit and the length of time they expect to be on the premises. This could allow the test facility to ensure that the appropriate personnel and documentation are available. In cases where particular documents or records are to be examined, it may be useful to identify these to the test facility in advance of the visit, so that they will be immediately available during the Test Facility Inspection.

Starting conference

PURPOSE: To inform the management and staff of the facility of the reason for the Test Facility Inspection or Study Audit that is about to take place, and to identify the facility areas, study(ies) selected for audit, documents and personnel likely to be involved.

The administrative and practical details of a Test Facility Inspection or Study Audit should be discussed with the management of the facility at the start of the visit. At the starting conference, Inspectors should:

- outline the purpose and scope of the visit;
- describe the documentation which will be required for the Test Facility Inspection, such as lists of on-going and completed studies, study plans, standard operating procedures, study reports, etc. Access to and, if necessary, arrangements for the copying of relevant documents should be agreed upon at this time;
- clarify or request information as to the management structure (organisation) and personnel of the facility;
- request information as to the conduct of studies not subject to GLP Principles in the areas of the test facility where GLP studies are being conducted;
- make an initial determination as to the parts of the facility to be covered during the Test Facility Inspection;
- describe the documents and specimens that will be needed for on-going or completed study(ies) selected for Study Audit;
- indicate that a closing conference will be held at the completion of the inspection.

Before proceeding further with a Test Facility Inspection, it is advisable for the Inspector(s) to establish contact with the facility's Quality Assurance (QA) unit.

As a general rule, when inspecting a facility, Inspectors will find it helpful to be accompanied by a member of the QA unit.

Inspectors may wish to request that a room be set aside for examination of documents and other activities.

Organisation and personnel

PURPOSE: To determine whether: the test facility has sufficient qualified personnel, staff resources and support services for the variety and number of studies undertaken; the organisational structure is appropriate; and management has established a policy regarding training and staff health surveillance appropriate to the studies undertaken in the facility.

The management should be asked to produce certain documents, such as:

- floor plans;
- facility management and scientific organisation charts;
- CVs of personnel involved in the type(s) of studies selected for the Study Audit;
- list(s) of on-going and completed studies with information on the type of study, initiation/completion dates, test system, method of application of test substance and name of Study Director;
- staff health surveillance policies;
- staff job descriptions and staff training programmes and records;

- an index to the facility's Standard Operating Procedures (SOPs);
- specific SOPs as related to the studies or procedures being inspected or audited;
- list(s) of the Study Directors and sponsors associated with the study(ies) being audited.

The Inspector should check, in particular:

- lists of on-going and completed studies to ascertain the level of work being undertaken by the test facility;
- the identity and qualifications of the Study Director(s), the head of the Quality Assurance unit and other personnel;
- existence of SOPs for all relevant areas of testing.

Quality Assurance Programme

PURPOSE: To determine whether the mechanisms used to assure management that studies are conducted in accordance with GLP Principles are adequate.

The head of the Quality Assurance (QA) unit should be asked to demonstrate the systems and methods for QA inspection and monitoring of studies, and the system for recording observations made during QA monitoring. Inspectors should check:

- the qualifications of the head of QA, and of all QA staff;
- that the QA unit functions independently from the staff involved in the studies;
- how the QA unit schedules and conducts inspections, how it monitors identified critical phases in a study, and what resources are available for QA inspections and monitoring activities;
- that, where studies are of such short duration that monitoring of each study is impracticable, arrangements exist for monitoring on a sample basis;
- the extent and depth of QA monitoring during the practical phases of the study;
- the extent and depth of QA monitoring of routine test facility operation;
- the QA procedures for checking the final report to ensure its agreement with the raw data;
- that management receives reports from QA concerning problems likely to affect the quality or integrity of a study;
- the actions taken by QA when deviations are found;
- the QA role, if any, if studies or parts of studies are done in contract laboratories;
- the part played, if any, by QA in the review, revision and updating of SOPs.

Facilities

PURPOSE: To determine if the test facility, whether indoor or outdoor, is of suitable size, design and location to meet the demands of the studies being undertaken.

The Inspector should check that:

- the design enables an adequate degree of separation so that, *e.g.*, test substances, animals, diets, pathological specimens, etc. of one study cannot be confused with those of another;
- environmental control and monitoring procedures exist and function adequately in critical areas, *e.g.* animal and other biological test systems rooms, test substance storage areas, laboratory areas;

- the general housekeeping is adequate for the various facilities and that there are, if necessary, pest control procedures.

Care, housing and containment of biological test systems

PURPOSE: To determine whether the test facility, if engaged in studies using animals or other biological test systems, has support facilities and conditions for their care, housing and containment, adequate to prevent stress and other problems which could affect the test system and hence the quality of data.

A test facility may be carrying out studies which require a diversity of animal or plant species, as well as microbial or other cellular or sub-cellular systems. The type of test systems being used will determine the aspects relating to care, housing or containment that the Inspector will monitor. Using his judgment, the Inspector will check, according to the test systems, that:

- there are facilities adequate for the test systems used and for testing needs;
- there are arrangements to quarantine animals and plants being introduced into the facility and that these arrangements are working satisfactorily;
- there are arrangements to isolate animals (or other elements of a test system, if necessary) known to be, or suspected of being, diseased or carriers of disease;
- there is adequate monitoring and record-keeping of health, behaviour or other aspects, as appropriate to the test system;
- the equipment for maintaining the environmental conditions required for each test system is adequate, well maintained, and effective;
- animal cages, racks, tanks and other containers, as well as accessory equipment, are kept sufficiently clean;
- analyses to check environmental conditions and support systems are carried out as required;
- facilities exist for removal and disposal of animal waste and refuse from the test systems and that these are operated so as to minimise vermin infestation, odours, disease hazards and environmental contamination;
- storage areas are provided for animal feed or equivalent materials for all test systems; that these areas are not used for the storage of other materials such as test substances, pest control chemicals or disinfectants, and that they are separate from areas in which animals are housed or other biological test systems are kept;
- stored feed and bedding are protected from deterioration by adverse environmental conditions, infestation or contamination.

Apparatus, materials, reagents and specimens

PURPOSE: To determine whether the test facility has suitably located, operational apparatus in sufficient quantity and of adequate capacity to meet the requirements of the tests being conducted in the facility and that the materials, reagents and specimens are properly labelled, used and stored.

The Inspector should check that:

- apparatus is clean and in good working order;
- records have been kept of operation, maintenance, verification, calibration and validation of measuring equipment and apparatus (including computerised systems);

- materials and chemical reagents are properly labelled and stored at appropriate temperatures and that expiry dates are not being ignored. Labels for reagents should indicate their source, identity and concentration and/or other pertinent information;
- specimens are well identified by test system, study, nature and date of collection;
- apparatus and materials used do not alter to any appreciable extent the test systems.

Test systems

PURPOSE: To determine whether adequate procedures exist for the handling and control of the variety of test systems required by the studies undertaken in the facility, *e.g.* chemical and physical systems, cellular and microbic systems, plants or animals.

Physical and chemical systems

The Inspector should check that:

- where required by study plans, the stability of test and reference substances was determined and that the reference substances specified in test plans were used;
- in automated systems, data generated as graphs, recorder traces or computer print-outs are documented as raw data and archived.

Biological test systems

Taking account of the relevant aspects referred to above relating to care, housing or containment of biological test systems, the Inspector should check that:

- test systems are as specified in study plans;
- test systems are adequately and, if necessary and appropriate, uniquely identified throughout the study; and that records exist regarding receipt of the test systems and document fully the number of test systems received, used, replaced or discarded;
- housing or containers of test systems are properly identified with all the necessary information;
- there is an adequate separation of studies being conducted on the same animal species (or the same biological test systems), but with different substances;
- there is an adequate separation of animal species (and other biological test systems), either in space or in time;
- the biological test system environment is as specified in the study plan or in SOPs for aspects such as temperature, or light/dark cycles;
- the recording of the receipt, handling, housing or containment, care and health evaluation is appropriate to the test systems;
- written records are kept of examination, quarantine, morbidity, mortality, behaviour, diagnosis and treatment of animal and plant test systems or other similar aspects as appropriate to each biological test system;
- there are provisions for the appropriate disposal of test systems at the end of tests.

Test and reference substances

PURPOSE: To determine whether the test facility has procedures designed i) to ensure that the identify, potency, quantity and composition of test and reference substances are in accordance with their specifications, and ii) to properly receive and store test and reference substances.

The Inspector should check that:

- there are written records on the receipt (including identification of the person responsible), and for the handling, sampling, usage and storage of tests and reference substances;
- test and reference substances containers are properly labelled;
- storage conditions are appropriate to preserve the concentration, purity and stability of the test and reference substances;
- there are written records on the determination of identity, purity, composition, stability, and for the prevention of contamination of test and reference substances, where applicable;
- there are procedures for the determination of the homogeneity and stability of mixtures containing test and reference substances, where applicable;
- containers holding mixtures (or dilutions) of the test and reference substances are labelled and that records are kept of the homogeneity and stability of their contents, where applicable;
- when the test is of longer than four weeks' duration, samples from each batch of test and reference substances have been taken for analytical purposes and that they have been retained for an appropriate time;
- procedures for mixing substances are designed to prevent errors in identification or cross-contamination.

Standard operating procedures

PURPOSE: To determine whether the test facility has written SOPs relating to all the important aspects of its operations, considering that one of the most important management techniques for controlling facility operations is the use of written SOPs. These relate directly to the routine elements of tests conducted by the test facility.

The Inspector should check that:

- each test facility area has immediately available relevant, authorised copies of SOPs;
- procedures exist for revision and updating of SOPs;
- any amendments or changes to SOPs have been authorised and dated;
- historical files of SOPs are maintained;
- SOPs are available for, but not necessarily limited to, the following activities:
    i)   receipt; determination of identity, purity, composition and stability; labelling; handling; sampling; usage; and storage of test and reference substances;
    ii) use, maintenance, cleaning, calibration and validation of measuring apparatus, computerised systems and environmental control equipment;
    iii) preparation of reagents and dosing formulations;
    iv) record-keeping, reporting, storage and retrieval of records and reports;
    v)   preparation and environmental control of areas containing the test systems;
    vi) receipt, transfer, location, characterisation, identification and care of test systems;

vii) handling of the test systems before, during and at the termination of the study;

viii) disposal of test systems;

ix) use of pest control and cleaning agents;

x) Quality Assurance Programme operations.

Performance of the study

PURPOSE: To verify that written study plans exist and that the plans and the conduct of the study are in accordance with GLP Principles.

The Inspector should check that:

- the study plan was signed by the Study Director;
- any amendments to the study plan were signed and dated by the Study Director;
- the date of the agreement to the study plan by the sponsor was recorded (where applicable);
- measurements, observations and examinations were in accordance with the study plan and relevant SOPs;
- the results of these measurements, observations and examinations were recorded directly, promptly, accurately and legibly and were signed (or initialled) and dated;
- any changes in the raw data, including data stored in computers, did not obscure previous entries, included the reason for the change and identified the person responsible for the change and the date it was made;
- computer-generated or stored data have been identified and that the procedures to protect them against unauthorised amendments or loss are adequate;
- the computerised systems used within the study are reliable, accurate and have been validated;
- any unforeseen events recorded in the raw data have been investigated and evaluated;
- the results presented in the reports of the study (interim or final) are consistent and complete and that they correctly reflect the raw data.

Reporting of study results

PURPOSE: To determine whether final reports are prepared in accordance with GLP Principles.

When examining a final report, the Inspector should check that:

- it is signed and dated by the Study Director to indicate acceptance of responsibility for the validity of the study and confirming that the study was conducted in accordance with GLP Principles;
- it is signed and dated by other principal scientists, if reports from co-operating disciplines are included;
- a Quality Assurance statement is included in the report and that it is signed and dated;
- any amendments were made by the responsible personnel;
- it lists the archive location of all samples, specimens and raw data.

Storage and retention of records

PURPOSE: To determine whether the facility has generated adequate records and reports and whether adequate provision has been made for the safe storage and retention of records and materials;

The Inspector should check:

- that a person has been identified as responsible for the archive;
- the archive facilities for the storage of study plans, raw data (including that from discontinued GLP Studies), final reports, samples and specimens and records of education and training of personnel;
- the procedures for retrieval of archived materials;
- the procedures whereby access to the archives is limited to authorised personnel and records are kept of personnel given access to raw data, slides, etc.;
- that an inventory is maintained of materials removed from, and returned to, the archives;
- that records and materials are retained for the required or appropriate period of time and are protected from loss or damage by fire, adverse environmental conditions, etc.

Study Audits

Test Facility Inspections will generally include, *inter alia*, Study Audits, which review on-going or completed studies. Specific Study Audits are also often requested by Regulatory Authorities, and can be conducted independently of Test Facility Inspections. Because of the wide variation in the types of studies which might be audited, only general guidance is appropriate, and Inspectors and others taking part in Study Audits will always need to exercise judgment as to the nature and extent of their examinations. The objective should be to reconstruct the study by comparing the final report with the study plan, relevant SOPs, raw data and other archived material.

In some cases, Inspectors may need assistance from other experts in order to conduct an effective Study Audit, *e.g.* where there is a need to examine tissue sections under the microscope.

When conducting a Study Audit, the Inspector should:

- obtain names, job descriptions and summaries of training and experience for selected personnel engaged in the study(ies) such as the Study Director and principal scientists;
- check that there is sufficient staff trained in relevant areas for the study(ies) undertaken;
- identify individual items of apparatus or special equipment used in the study and examine the calibration, maintenance and service records for the equipment;
- review the records relating to the stability of the test substances, analyses of test substance and formulations, analyses of feed, etc.;
- attempt to determine, through the interview process if possible, the work assignments of selected individuals participating in the study to ascertain if these individuals had the time to accomplish the tasks specified in the study plan or report;
- obtain copies of all documentation concerning control procedures or forming integral parts of the study, including:
  i) the study plan;
  ii) OPs in use at the time the study was done;

iii) log books, laboratory notebooks, files, worksheets, print-outs of computer-stored data, etc.; check calculations, where appropriate;

iv) the final report.

In studies in which animals (i.e., rodents and other mammals) are used, the Inspectors should follow a certain percentage of individual animals from their arrival at the test facility to autopsy. They should pay particular attention to the records relating to:

- animal body weight, food/water intake, dose formulation and administration, etc.;
- clinical observations and autopsy findings;
- clinical chemistry;
- pathology.

Completion of Inspection or Study Audit

When a Test Facility Inspection or Study Audit has been completed, the Inspector should be prepared to discuss his findings with representatives of the test facility at a Closing Conference and should prepare a written report, i.e. the Inspection Report.

A Test Facility Inspection of any large facility is likely to reveal a number of minor deviations from GLP Principles, but, normally, these will not be sufficiently serious to affect the validity of studies emanating from that test facility. In such cases, it is reasonable for an Inspector to report that the facility is operating in compliance with GLP Principles according to the criteria established by the (National) GLP Monitoring Authority. Nevertheless, details of the inadequacies or faults detected should be provided to the test facility and assurances sought from its senior management that action will be taken to remedy them. The Inspector may need to revisit the facility after a period of time to verify that necessary action has been taken.

If a serious deviation from the GLP Principles is identified during a Test Facility Inspection or Study Audit which, in the opinion of the Inspector, may have affected the validity of that study, or of other studies performed at the facility, the Inspector should report back to the (National) GLP Monitoring Authority. The action taken by that authority and/or the regulatory authority, as appropriate, will depend upon the nature and extent of the non-compliance and the legal and/or administrative provisions within the GLP Compliance Programme.

Where a Study Audit has been conducted at the request of a Regulatory Authority, a full report of the findings should be prepared and sent via the relevant (National) GLP Monitoring Authority to the Regulatory Authority concerned.

3. GUIDANCE FOR THE PREPARATION OF GLP INSPECTION REPORTS*

History

Under the auspices of the OECD Panel on Good Laboratory Practice, a working group met in Rockville, Maryland, from 21 through 23 September 1994, to develop harmonised guidance for the preparation of GLP inspection reports. The working group was chaired by Mr. Paul Lepore of the United States Food and Drug Administration. Participants were from National GLP Compliance Monitoring Authorities in the following countries: Canada, France, Germany, Norway, Sweden, Switzerland and the USA. The working group reached consensus on a draft document aimed at providing guidance for GLP monitoring authorities on the information on specific Test Facility Inspections to be exchanged with their colleagues in other GLP monitoring authorities.

The Panel on GLP reviewed and amended the draft document prepared by the working group and subsequently forwarded the document to the Joint Meeting of the Chemicals Group and Management Committee of the Special Programme on the Control of Chemicals for endorsement in 1995. It is derestricted under the authority of the Secretary-General.

Introduction

One of the goals of the work of the OECD Panel on Good Laboratory Practice is to facilitate the sharing of information from GLP compliance monitoring programmes conducted by member countries. This goal requires more than the promulgation of enforceable principles of GLP and the conduct of an inspection programme by the national monitoring authority. It is also necessary to have the reports of the inspections prepared in a useful and consistent manner. The Guidance for the Preparation of GLP Inspection Reports developed by the Panel on GLP set forth below suggests elements and/or concepts that can contribute to a useful report of a GLP inspection and study audit. It may be used by member countries as a component of their compliance monitoring programme.

Report headings

There are many acceptable ways to organise an inspection report, but the key is to make sure that it contains the required information and meets the requirements of the regulatory authority. Generally, report headings include a Summary, an Introduction, a Narrative, a Summary of the Exit Discussion, and Annexes. All of the information presented under these headings should portray an accurate picture of the adherence of the testing facility to the Principles of GLP and the quality of any study report that may have been audited.

---

* No. 9 in the OECD Series on Principles of Good Laboratory Practice and Compliance Monitoring.

The narrative headings may contain information as follows:

1. **Summary**

    The summary section of the report should be presented first and should provide background information on the test facility, the type of inspection that was conducted, the deviations from the GLP Principles that were noted, and the responses of the test facility to the presented deviations. In accord with national practice, the report may include the compliance designation of the laboratory that was assigned by the inspectors.

2. **Introduction**

    The introductory section should include some or all of the following elements:

    2.1. The purpose and general description of the inspection, including the legal authority of the inspectors and the quality standards serving as the basis for the inspection.

    2.2. An identification of the inspectors and the dates of inspection.

    2.3. A description of the type of inspection (facility, study audit, etc.).

    2.4. An identification of the test facility, including corporate identity, postal address, and contact person(s) [with telephone and telefax number(s)].

    2.5. A description of the test facility identifying the categories of test substances and testing that is done and presenting information on the physical layout and the personnel.

    2.6. The date of the previous GLP inspection, resulting GLP compliance status, and any relevant changes made by the test facility since that inspection.

3. **Narrative**

    The Narrative portion of the report should contain a complete and factual description of the observations made and activities undertaken during the course of the inspection. Generally, the information recorded in this section should be reflected under the headings in the GLP Principles, as listed below:

    3.1. Organisation and personnel.

    3.2. Quality Assurance Programme.

    3.3. Facilities.

    3.4. Apparatus, materials, reagents and specimens.

    3.5. Test systems.

    3.6. Test and reference substances.

    3.7. Standard operating procedures.

    3.8. Performance of the study.

    3.9. Reporting of study results.

    3.10. Storage and retention of records.

    Deviations from the GLP Principles should be supported by documentation (i.e. photocopies, photographs, test samples, etc.). All such documentation should be referenced and discussed in the Narrative and attached in the Annexes.

    When a study has been selected for audit, the inspection report should describe the procedure for conducting the audit, including a description of the portion of the data or study that was actually examined. Any findings during the audit should be described in the Narrative and documented in the Annexes.

4. **Exit discussion**

At the end of an inspection/study audit, an Exit Conference should be held between the inspection team and the responsible management of the test facility, at which GLP deviations found during the inspection/study audit may be discussed. During this Exit Conference, if allowed by national policy, a written list of observations should be presented describing the GLP deviations if any have been observed. The exit discussion should be summarized in this section.

The report should note the date and time of the Exit Conference; the names of attendees (inspection team, facility and others), with their affiliations. It should also give a brief summary of GLP deviations noted by the inspection team during the facility inspection and/or Study Audits. Responses of facility representatives to the inspection team's remarks should also be described.

In the case where a written list of observations has been made available, the test facility should acknowledge the inspectors' findings and make a commitment to take corrective action.

If a receipt of documents taken by the inspection team was prepared and signed by facility management, the person to whom the receipt for documents was provided should be identified. A copy of the receipt should be included in the Annexes.

5. **Annexes**

The Annexes should contain copies of documents that have been referenced in the report. Such documents may include:

- Organisational charts of the facility.
- The agenda for the inspection.
- A listing of SOPs that have been demonstrated during the inspection.
- A listing of deviations that have been observed.
- Photocopies that document observed deviations.

Other information

In addition to the information described above, reports may contain other headings and information, as appropriate or as required by a member country's compliance monitoring programme. For example, the inspection report may address correction of deficiencies noted during previous inspections or any corrective action taken during the current inspection. Others may include a cover page, which contains descriptive information that briefly identifies the inspection. Others find it useful to use a table of contents, especially when the inspection is of a large, complex facility to categorise, index, and identify information in the report. Some reports include a "conclusion" section which notifies the testing facility of the compliance status classification as judged by the inspection. Any, or all of these, are acceptable.

Approval

Reports should be signed and dated by the lead inspector and by other inspectors in accordance with their responsibilities.

PART III

Chapter 4

Advisory Document of the Working Group on GLP

1. REQUESTING AND CARRYING OUT INSPECTIONS AND Study Audits IN ANOTHER COUNTRY*

History

Environmental health and safety studies for the assessment of chemicals and chemical products are increasingly being carried out in multiple sites. This holds not only for field studies, but also for various phases of toxicology studies. The Revised Principles of Good Laboratory Practice, adopted by OECD in 1997, cover the various aspects of the organisation of such studies. Nevertheless, the Working Group on Good Laboratory Practice felt that further guidance was needed about requesting and carrying out inspections and Study Audits of multi-site studies when the study site(s) are located in another country than that of the main test facility, as accorded by the 1989 Council Decision-Recommendation on Compliance with Principles of GLP [C(89)87(Final), Part II, 2.iii)].

The Working Group therefore established a Steering Group on Multi-site Studies under the leadership of Germany. The Group met in Berlin on 2-3 September 1999 and included participants from the following countries: Denmark, France, Germany, the Netherlands, Sweden, Switzerland, the United Kingdom and the United States. It was chaired by Hans-Wilhelm Hembeck (Germany). The document prepared by the Steering Group was examined by the Working Group at its 12th Meeting in January 2000, where it was amended and endorsed.

The Joint Meeting of the Chemicals Committee and the Working Party on Chemicals, Pesticides and Biotechnology endorsed the document in 2000. It is declassified under the authority of the Secretary-General.

Introduction

In the 1989 Council Decision-Recommendation on Compliance with the Principles of Good Laboratory Practice [C(89)87/Final], member countries decided that, for purposes of the recognition of the assurance by another member country that test data have been generated in accordance with GLP Principles, countries "shall implement procedures whereby, where good reason exists, information concerning GLP compliance of a test facility (including information focusing on a particular study) within their jurisdiction can be sought by another member country." It is understood that such procedures should only be applied in exceptional circumstances.

The Working Group on Good Laboratory Practice proposed clarification of this decision based on the Revised OECD Principles of GLP and recommended the procedures set out below. This clarification was considered necessary, since it was recognised that some test facilities have test sites located under the jurisdiction of another country. These facilities or sites may not necessarily be part of the GLP compliance monitoring programme

---

* No. 12 in the OECD Series on Principles of Good Laboratory Practice and Compliance Monitoring.

of the country of location, although many member countries consider this desirable and useful.

The Working Group agreed that the use of the term "test facility" in the 1989 Council Act encompassed both "test facility" and "test site" as defined in the Revised OECD Principles of GLP. Therefore, any member country can request an inspection/study audit from both test facilities and test sites located in another country. This request could concern any organisation associated with regulated GLP studies, whether these be main test facilities or test sites (dependent or independent of the test facility) which carry out phases of a study such as chemical analysis, histopathology or field studies.

Requests can also be made to inspect associated organisations such as independent Quality Assurance or archiving facilities, if national legislation allows. However, this information exchange could be of a more informal nature and such operations need not necessarily appear in the annual overviews of inspected facilities exchanged among members of the Working Group on GLP. These annual overviews should, however, include test facilities and test sites which were inspected or in which Study Audits were carried out.

In order to "implement procedures" to allow for this information exchange to take place smoothly and efficiently among monitoring authorities, to avoid duplication and wasting of resources and to assure that there is adequate compliance monitoring, the Working Group agreed that a process needed to be established for requesting inspections or Study Audits in another country.

The Working Group agreed that, if justifiable requests to confirm compliance with GLP are made, every effort should be made to accommodate requests for inspections or Study Audits of test facilities or sites in other countries. If the country where the facility or site is located cannot accommodate the request in the framework of its current GLP monitoring programme and/or schedule, an alternative could be to allow the requesting country to undertake the inspection and/or audit itself (at its own expense, as mutually agreed by both parties). Refusal to accommodate such requests may result in rejection of studies from the facility or site concerned. It was agreed that all members of the Working Group on GLP should be informed of such refusals and that the circumstances should be discussed in the Working Group.

Recommended procedures to be followed in requesting and carrying out inspections and Study Audits in another country

1. The request for an inspection and/or study audit in another country should be made in writing and justified. The two countries should work out the arrangements to accommodate the request and for provision appropriate materials in a timely manner.
2. The liaison and lines of communication should be between the two National GLP Monitoring Authorities concerned.
3. The inspection/study audit will normally be led by the monitoring authority where the facility and/or site is located. An inspector or inspectors from the requesting country can be present at the inspection/study audit. Receiving authorities may participate if appropriate. The requesting country shall cover any costs involved for its own personnel.

4. The inspection/study audit report should be submitted to the requesting country (in an appropriate language as agreed between the two countries), with the appropriate measures taken to cover concerns about protection of commercial and industrial secrecy, as required by national legislation.

5. Any major findings during such inspections/Study Audits should be followed up by the appropriate monitoring authority(ies).

6. Financial arrangements for inspections and Study Audits undertaken in this context will be made by the country in which they take place. The requesting country cannot be charged for this work.

7. Inspections and Study Audits undertaken in this context should appear in the annual overview of the country that led the inspection/study audit.

ANNEX

OECD Council Acts Related to the Mutual Acceptance of Data

Introduction: Legally binding agreements in OECD on Mutual Acceptance of Data in the Assessment of Chemicals (MAD)

The OECD is not a supranational organisation, but rather a forum for discussion where governments express their points of view, share their experiences and search for common ground. If member countries consider it appropriate, an accord can be embodied in a formal OECD Council Act, which is agreed at the highest level of OECD, the Council.

In general, there are two types of Council Act. A Council Decision, which is legally binding on OECD member countries, and a Council Recommendation, which is a strong expression of political will. In the area of chemicals, there are three Council Decisions relating to the Mutual Acceptance of Data (MAD), which are a cornerstone of the work on chemicals management in OECD.

The testing of chemicals is labour-intensive and expensive, and testing the same chemical in several countries adds to the cost in time, resources and laboratory animals. Because of the need to relieve some of this burden, the OECD Council adopted a Decision in 1981 stating that "data generated in a member country in accordance with OECD Test Guidelines and Principles of Good Laboratory Practice (GLP) shall be accepted in other member countries for assessment purposes and other uses relating to the protection of human health and the environment." The 1981 Council Decision sets the policy context agreed by all OECD member countries which established that safety data developed in one member country will be accepted for use by the relevant registration authorities in assessing the chemical or product in another OECD country, *i.e.* the data does not have to be generated a second time for the purposes of safety assessment.

A further Council Act was adopted in 1989 to provide safeguards for assurance that the data is indeed developed in compliance with the Principles of GLP. This Council Decision-Recommendation on Compliance with GLP establishes procedures for monitoring GLP compliance through government inspections and study audits, as well as a framework for international liaison among monitoring and data-receiving authorities. A 1997 Council Decision on the Adherence of Non-Member Countries to the Council Acts related to the Mutual Acceptance of Data in the Assessment of Chemicals sets out a step-wise procedure for non-OECD countries with a significant chemical industry input to take part as full members in this system.

The three Council Acts are reproduced on the following pages.

ANNEX

1. DECISION OF THE COUNCIL
Concerning the Mutual Acceptance of Data in the Assessment of Chemicals
[C(81)30(Final)]

(Adopted by the Council at its 535th Meeting on 12 May 1981)

The Council,

Having regard to Articles 2(a), 2(d), 3, 5(a) and 5(b) of the Convention on the Organisation for Economic Co-operation and Development of 14 December 1960;

Having regard to the Recommendation of the Council of 26 May 1972, on Guiding Principles concerning International Economic Aspects of Environmental Policies [C(72)128];

Having regard to the Recommendation of the Council of 14 November 1974, on the Assessment of the Potential Environmental Effects of Chemicals [C(74)215];

Having regard to the Recommendation of the Council of 26 August 1976, concerning Safety Controls over Cosmetics and Household Products [C(76)144(Final)];

Having regard to the Recommendation of the Council of 7 July 1977, establishing Guidelines in respect of Procedure and Requirements for Anticipating the Effects of Chemicals on Man and in the Environment [C(77)97(Final)];

Having regard to the Decision of the Council of 21 September 1978, concerning a Special Programme on the Control of Chemicals and the Programme of Work established therein [C(78)127(Final)];

Having regard to the Conclusions of the First High Level Meeting of the Chemicals Group of 19 May 1980, dealing with the control of health and environmental effects of chemicals [ENV/CHEM/HLM/80.M/1];

Considering the need for concerted action amongst OECD member countries to protect man and his environment from exposure to hazardous chemicals;

Considering the importance of international production and trade in chemicals and the mutual economic and trade advantages which accrue to OECD member countries from harmonisation of policies for chemicals control;

Considering the need to minimise the cost burden associated with testing chemicals and the need to utilise more effectively scarce test facilities and specialist manpower in member countries;

Considering the need to encourage the generation of valid and high quality test data and noting the significant actions taken in this regard by OECD member countries through provisional application of OECD Test Guidelines and OECD Principles of Good Laboratory Practice;

Considering the need for and benefits of mutual acceptance in OECD countries of test data used in the assessment of chemicals and other uses relating to protection of man and the environment;

On the proposal of the High Level Meeting of the Chemicals Group, endorsed by the Environment Committee;

PART I

1. DECIDES that data generated in the testing of chemicals in an OECD member country in accordance with OECD Test Guidelines and OECD Principles of Good Laboratory Practice shall be accepted in other member countries for purposes of assessment and other uses relating to the protection of man and the environment.

2. DECIDES that, for the purposes of this decision and other Council actions the terms OECD Test Guidelines and OECD Principles of Good Laboratory Practice shall mean guidelines and principles adopted by the Council.

3. INSTRUCTS the Environment Committee to review action taken by member countries in pursuance of this Decision and to report periodically thereon to the Council.

4. INSTRUCTS the Environment Committee to pursue a programme of work designed to facilitate implementation of this Decision with a view to establishing further agreement on assessment and control of chemicals within Member countries.

PART II

To implement the Decision set forth in Part I:

1. RECOMMENDS that member countries, in the testing of chemicals, apply the OECD Test Guidelines and the OECD Principles of Good Laboratory Practice, set forth respectively in Annexes I and II* which are integral parts of this text.

2. INSTRUCTS the Management Committee of the Special Programme on the Control of Chemicals in conjunction with the Chemicals Group of the Environment Committee to establish an updating mechanism to ensure that the aforementioned test guidelines are modified from time to time, as required through the revision of existing Guidelines or the development of new Guidelines.

3. INSTRUCTS the Management Committee of the Special Programme on the Control of Chemicals to pursue its programme of work in such a manner as to facilitate internationally-harmonised approaches to assuring compliance with the OECD Principles of Good Laboratory Practice and to report periodically thereon to the Council.

---

* Annex I to the Council Decision (the OECD Test Guidelines) was published separately and is regularly updated. Annex II (the OECD Principles of Good Laboratory Practice) was revised in 1997 and can be found in Part I of this publication.

2. COUNCIL DECISION-RECOMMENDATION
on Compliance with Principles of Good Laboratory Practice
[C(89)87(Final)]

(Adopted by the Council at its 717th Meeting on 2 October 1989)

The Council,

Having regard to Articles 5a) and 5b) of the Convention on the Organisation for Economic Co-operation and Development of 14 December 1960;

Having regard to the Recommendation of the Council of 7th July, 1977 Establishing Guidelines in Respect of Procedure and Requirements for Anticipating the Effects of Chemicals on Man and in the Environment [C(77)97(Final)];

Having regard to the Decision of the Council of 12 May 1981 concerning the Mutual Acceptance of Data in the Assessment of Chemicals [C(81)30(Final)] and, in particular, the Recommendation that member countries, in the testing of chemicals, apply the OECD Principles of Good Laboratory Practice, set forth in Annex 2 of that Decision;

Having regard to the Recommendation of the Council of 26 July 1983 concerning the Mutual Recognition of Compliance with Good Laboratory [C(83)95(Final)];

Having regard to the conclusions of the Third High Level Meeting of the Chemicals Group (OECD, Paris, 1988);

Considering the need to ensure that test data on chemicals provided to regulatory authorities for purposes of assessment and other uses related to the protection of human health and the environment are of high quality, valid and reliable;

Considering the need to minimise duplicative testing of chemicals, and thereby to utilise more effectively scarce test facilities and specialist manpower, and to reduce the number of animals used in testing;

Considering that recognition of procedures for monitoring compliance with good laboratory practice will facilitate mutual acceptance of data and thereby reduce duplicative testing of chemicals;

Considering that a basis for recognition of compliance monitoring procedures is an understanding of, and confidence in, the procedures in the member country where the data are generated;

Considering that harmonised approaches to procedures for monitoring compliance with good laboratory practice would greatly facilitate the development of the necessary confidence in other countries' procedures;

On the proposal of the Joint Meeting of the Management Committee of the Special Programme on the Control of Chemicals and the Chemicals Group, endorsed by the Environment Committee.

PART I

GLP Principles and Compliance Monitoring

1. DECIDES that member countries in which testing of chemicals for purposes of assessment related to the protection of health and the environment is being carried out pursuant to principles of good laboratory practice that are consistent with the OECD Principles of Good Laboratory Practice as set out in Annex 2 of the Council Decision [C(81)30(Final)] (hereafter called "GLP Principles") shall:

   i) establish national procedures for monitoring compliance with GLP Principles, based on laboratory inspections and Study Audits;

   ii) designate an authority or authorities to discharge the functions required by the procedures for monitoring compliance; and

   iii) require that the management of test facilities issue a declaration, where applicable, that a study was carried out in accordance with GLP Principles and pursuant to any other provisions established by national legislation or administrative procedures dealing with good laboratory practice.

2. RECOMMENDS that, in developing and implementing national procedures for monitoring compliance with GLP Principles, member countries apply the "Guides for Compliance Monitoring Procedures for Good Laboratory Practice" and the "Guidance for the Conduct of Laboratory Inspections and Study Audits", set out respectively in Annexes I and II which are an integral part of this Decision-Recommendation.

PART II

Recognition of GLP Compliance among Member Countries

1. DECIDES that member countries shall recognise the assurance by another member country that test data have been generated in accordance with GLP Principles, if such other member country complies with Part I above and Part II paragraph 2 below.

2. DECIDES that, for purposes of the recognition of the assurance in paragraph 1 above, member countries shall:

   i) designate an authority or authorities for international liaison and for discharging other functions relevant to the recognition as set out in this Part and in the Annexes to this Decision-Recommendation;

   ii) exchange with other member countries relevant information concerning their procedures for monitoring compliance, in accordance with the guidance set out in Annex III which is an integral part of this Decision-Recommendation, and

   iii) implement procedures whereby, where good reason exists, information concerning GLP compliance of a test facility (including information focusing on a particular study) within their jurisdiction can be sought by another member country.

3. DECIDES that the Council Recommendation concerning the Mutual Recognition of Compliance with Good Laboratory Practice [C(83)95(Final)] shall be repealed.

PART III
Future OECD Activities

1. INSTRUCTS the Environment Committee and the Management Committee of the Special Programme on the Control of Chemicals to ensure that the "Guides for Compliance Monitoring Procedures for Good Laboratory Practice" and the "Guidance for the Conduct of Laboratory Inspections and Study Audits" set out in Annexes I and II are updated and expanded, as necessary, in light of developments and experience of member countries and relevant work in other international organisations.

2. INSTRUCTS the Environment Committee and the Management Committee of the Special Programme on the Control of Chemicals to pursue a programme of work designed to facilitate the implementation of this Decision-Recommendation, and to ensure continuing exchange of information and experience on technical and administrative matters related to the application of GLP Principles and the implementation of procedures for monitoring compliance with good laboratory practice.

3. INSTRUCTS the Environment Committee and the Management Committee of the Special Programme on the Control of Chemicals to review actions taken by member countries in pursuance of this Decision-Recommendation.

Annex I to C(89)87(Final)/Revised in C(95)8(Final)
GUIDES FOR COMPLIANCE MONITORING PROCEDURES FOR GOOD LABORATORY PRACTICE

See Part III, Chapter 1 of this publication.

* * *

Annex II to C(89)87(Final)/Revised in C(95)8(Final)
GUIDANCE FOR THE CONDUCT OF LABORATORY INSPECTIONS AND Study Audits

See Part III, Chapter 2 of this publication.

* * *

Annex III to C(89)87(Final)/Revised in C(95)8(Final)
GUIDANCE FOR THE EXCHANGE OF INFORMATION CONCERNING NATIONAL PROGRAMMES FOR MONITORING OF COMPLIANCE WITH PRINCIPLES OF GOOD LABORATORY PRACTICES

Part II, paragraph 2 of the Council Act contains a Decision that member countries exchange information related to their programmes for monitoring of compliance with GLP Principles. This Annex provides guidance concerning the types of information which should be exchanged. While information concerning all of the aspects covered in the "Guides for Compliance Monitoring Programmes procedures for Good Laboratory Practic" (Annex I) are relevant to an understanding of other member countries' programmes for GLP Compliance Monitoring, certain types of information are of particular importance. These include:

- the GLP Principles adopted nationally;
- the scope of the national programme for monitoring compliance with GLP Principles in terms of the types of chemicals and tests covered;
- the identity, legal status, and organisational structure of the (National) GLP Monitoring Authority(ies);
- the procedures followed during Test Facility Inspections and Study Audits, and the periodicity of inspections and/or criteria for inspection schedules;
- the number and qualifications of Inspectors;
- the actions available to the (National) GLP Monitoring Authority(ies) in cases of non-compliance, including the ability to inform other member countries, when necessary, of the results of Test Facility Inspections and Study Audits;
- the arrangements for protecting confidentiality of information;
- the procedures for initiating, conducting and reporting on Test Facility Inspections and Study Audits at the request of other member countries;
- the procedures for obtaining information on test facilities which have been inspected by a (National) GLP Monitoring Authority of another member country, including such facilities' compliance status; and
- the nature of test facility certifications that studies were carried out following GLP Principles.

Where serious deviations which may have affected specific studies are found, the (National) GLP Monitoring Authority should consider the need to inform relevant (National) GLP Monitoring Authorities in other member countries of their findings.

The names of test facilities subject to Test Facility Inspections within a (National) GLP Compliance Programme, their levels of compliance with the national GLP Principles and the date(s) the inspections were conducted should be made available annually to (National) GLP Monitoring Authorities in other member countries upon request (see "Guidance for GLP Monitoring Authorities for the Preparation of Annual Overviews of Test Facilities Inspected" set out in the Appendix to this Annex.)

Recognition of national programmes for monitoring compliance with GLP Principles may not be immediately forthcoming from other member countries. Member countries should be prepared to meet genuine concerns in a co-operative way. It may be that a member country is unable to judge the acceptability of the GLP Compliance Monitoring programmes of another solely on the basis of the exchange of written information. In such cases, member countries may seek the assurance they require through consultation and discussion with relevant (National) GLP Monitoring Authorities. In this context, OECD provides a forum for the discussion and solving of problems relating to the international harmonisation and acceptance of GLP Compliance Monitoring programmes.

To facilitate international liaison and the continuing exchange of information, the establishment of a single GLP Monitoring Authority covering all good laboratory practice activities within a member country has obvious advantages. Where more than one authority exists, a member country should ensure that they operate in a consistent way, and have similar GLP Compliance Programmes. The authority or authorities with responsibilities for international contacts should be identified by member countries.

Situations will arise where a national Regulatory Authority of a member country will need to request information on the GLP Compliance Status of a test facility located in another member country. On rare occasions, and where good reason exists, a particular Study Audit may be requested by a Regulatory Authority of another member country. Arrangements should be provided whereby these requests may be fulfilled and the results reported back to the requesting Regulatory Authority.

Formal international contact should be established for the exchange of information between GLP Monitoring Authorities. However, this should not be understood to prevent informal contacts between Regulatory Authorities and the GLP Monitoring Authority in another member country, to the extent that such contacts are accepted by the member countries concerned.

National authorities should note that authorities from another member country may wish to be present at a Test Facility Inspection or Study Audit that they have specifically requested; or they may wish that representative(s) from the member country seeking a special Test Facility Inspection or Study Audit be present at that inspection or audit. In these cases, member countries should enable Inspectors from another member country to participate in facility inspections and Study Audits carried out by their GLP Monitoring Authority.

Appendix to Annex III to C(89)87(Final)/Revised in C(95)8(Final)

GUIDANCE FOR GOOD LABORATORY PRACTICE MONITORING AUTHORITIES FOR THE PREPARATION OF ANNUAL OVERVIEWS OF TEST FACILITIES INSPECTED

Overviews of GLP inspections should be circulated to members of the OECD Panel on GLP and the OECD Secretariat annually before the end of March. The following minimum set of information should allow harmonisation of the overviews exchanged among National GLP Monitoring Authorities:

1. **Identification of the facility inspected:** Sufficient information should be included to make the identification of the facility unequivocal, *i.e.* the name of the test facility, the city and country in which it is located, including inspections abroad.

2. **Dates of inspections and decisions:** month and year of inspection, and, if appropriate, date of final decision on GLP compliance status.

3. **Nature of inspection:** A clear indication should be given of whether a full GLP inspection or only a study audit was carried out, as well as whether the inspection was routine or not and any other authorities which were involved.

4. **Areas of expertise of the facility inspected:** Since GLP compliance is related to the tests performed by a facility, the area(s) of expertise of the test facilities inspected should be included in the annual overviews, using the following broad categories:

   1. physical-chemical testing;
   2. toxicity studies;
   3. mutagenicity studies;
   4. environmental toxicity studies on aquatic and terrestrial organisms;
   5. studies on behaviour in water, soil and air; bioaccumulation;
   6. residue studies;
   7. studies on effects on mesocosms and natural ecosystems;
   8. analytical and clinical chemistry testing;
   9. other studies, specify.

   It is emphasised that these categories are to be used in a flexible manner on a case-by-case basis and that the aim is to provide information related to GLP compliance of test facilities that will be useful for other national monitoring authorities.

5. **Compliance status:** The three following categories should be used to report the compliance status of facilities:

   - in compliance;
   - not in compliance;
   - pending (with explanation).

   In light of the fact that "pending" is interpreted differently by member countries and that the varying legal and administrative systems do not allow for harmonised use of the term, explanations must accompany the use of the "pending" status in the national overview of test facilities inspected. Such explanations could include, *e.g.*, "pending reinspection", "pending responses from test facility", "pending completion of administrative procedures", etc.

6. **Comments:** If appropriate, further comments can be made.

7. **Major deficiencies:** At a minimum, individual studies for which a study audit has revealed serious GLP deficiencies and which have consequently been rejected by receiving authorities should be reported in the annual overviews of test facilities inspected. Since many studies are submitted to authorities in several countries at the same time, however, it is recommended that this kind of information be circulated among national authorities as rapidly as possible on an *ad hoc* basis, when necessary in addition to the annual overviews.

8. **Statements of compliance:** When statements of compliance are provided to facilities by national monitoring authorities, they should use the same terminology and categories as the annual overviews.

9. **Circulation of annual overviews:** Overviews should be circulated annually before the end of March to the members of the GLP Panel and the OECD Secretariat. This information can be released to the public on request.

3. COUNCIL DECISION
concerning the Adherence of Non-member Countries to the Council Acts related to the Mutual Acceptance of Data in the Assessment of Chemicals [C(81)30(Final) AND C(89)87(Final)] [C(97)114/Final]

(Adopted by the Council at its 912th meeting on 26 November 1997)

The Council,

Having regard to Articles 5(a) and 5(c) of the Convention on the Organisation for Economic Co-operation and Development of 14 December 1960;

Having regard to the Decision of the Council of 12 May 1981, concerning the Mutual Acceptance of Data in the Assessment of Chemicals [C(81)30(Final)];

Having regard to the Decision of the Council of 26 July 1983, concerning the Protection of Proprietary Rights to Data submitted in Notification of New Chemicals [C(83)96(Final)] and the Recommendations of the same date concerning the Exchange of Confidential Data on Chemicals [C(83)97(Final)] and the OECD List of Non-Confidential Data on Chemicals [C(83)98(Final)];

Having regard to the Decision Recommendation of the Council of 2 October 1989 on Compliance with Principles of Good Laboratory Practice [C(89)87(Final)] as amended];

Considering that effective implementation of the OECD Council Acts [C(81)30(Final)] and [C(89)87(Final)] is essential in view of the extension of these acts to adherence by non-member countries;

Recognising that the conclusion of agreements among members and with non-member countries constitutes a means for effective implementation of these Council Acts;

Recognising that adherence to the OECD Council Acts does not preclude use or acceptance of test data obtained in accordance with other scientifically valid and specified test methods, as developed for specific chemical product areas;

Considering that, on 14 June 1992 the United Nations Conference on Environment and Development, in Chapter 19, section E of Agenda 21, recommended that governments and international organisations should co-operate, particularly with developing countries, to develop appropriate tools for management of chemicals;

Considering the commitments made by Ministers at the meeting of the Council at Ministerial level of 23 and 24 May 1995 to support the integration of developing countries and economies in transition into the world economic system, and to pursue further progress toward a better environment;

Considering that member countries and non-member countries would derive both economic and environmental benefits from enlarged participation in the OECD Council Acts related to mutual acceptance of data in the assessment of chemicals;

Considering that non-member countries are increasingly demonstrating an interest in participating in the OECD Council Acts related to mutual acceptance of data in the assessment of chemicals;

Considering that the chemical industries in all nations have an interest in harmonised testing requirements and will benefit from the elimination of costly, duplicative testing and the avoidance of non-tariff barriers to trade;

Considering that expanded international co-operation to reduce duplicative testing would, in the process, diminish the use of animals for safety testing;

Considering, therefore, that it is appropriate and timely to pursue broadened international participation in the OECD programme on mutual acceptance of data in the assessment of chemicals, specifically by opening up the relevant OECD Council Acts to adherence by non-member countries and that a clear administrative procedure is required to facilitate this process;

On the proposal of the Joint Meeting of the Chemicals Group and Management Committee of the Special Programme on the Control of Chemicals, endorsed by the Environment Policy Committee;

1. DECIDES to open the OECD Council Acts related to the mutual acceptance of data in the assessment of chemicals* to adherence by non-member countries which express their willingness and demonstrate their ability to participate therein.

2. DECIDES that non-member countries adhering to the Council Acts shall be entitled to join the part of the OECD Chemicals Programme involving the mutual acceptance of data, with the same rights and obligations as member countries.

3. DECIDES that adherence to the Council Acts and participation in the part of the OECD Chemicals Programme related to the mutual acceptance of data shall be governed by the procedure set out in the Appendix to this Decision, of which it forms an integral part.

4. RECOMMENDS that member countries, with a view to facilitating the extension of the Council Acts to non-member countries, take or pursue all available means to ensure the most effective implementation of the Council Acts. Pending this effective implementation of the Council Acts by non-members, member countries shall be free to establish mutual acceptance of data with non-member countries on a bilateral basis.

5. INSTRUCTS the Management Committee of the Special Programme on the Control of Chemicals to assume responsibility for promoting international awareness of the Council Acts, with a view to informing, advising and otherwise encouraging non-member countries to participate in the programmes and activities that have been established by OECD countries pursuant to these Council Acts. Further, the Management Committee should monitor closely the technical aspects of implementation of the procedure set out in the Appendix, review the implementation of this Decision, and report thereon to Council within three years.

---

* These Council Acts are: the 1981 Council Decision concerning the Mutual Acceptance of Data in the Assessment of Chemicals [C(81)30(Final) as amended], together with the OECD Guidelines for the Testing of Chemicals and the OECD Principles of Good Laboratory Practice, and the 1989 Council Decision-Recommendation on Compliance with Principles of Good Laboratory Practice [C(89)87(Final) as amended] and are hereafter referred to as "the Council Acts".

Annex to C(97)114(Final)

PROCEDURE FOR ADHERENCE OF NON-MEMBER COUNTRIES TO THE COUNCIL ACTS RELATED TO THE MUTUAL ACCEPTANCE OF DATA IN THE ASSESSMENT OF CHEMICALS

i) The OECD Secretariat should ensure that an interested non-member country is provided with full information on the rights and obligations associated with adhering to the OECD Council Acts related to mutual acceptance of data in the assessment of chemicals.

ii) At the invitation of the Council, the interested non-member country would confirm, at an appropriate level, that it would agree to provisionally adhere to the Council Acts and to accept, for purposes of assessment and other uses relating to the protection of man and environment, data generated in the testing of chemicals with OECD Test Guidelines and OECD Principles of Good Laboratory Practice.

iii) Following such invitation, confirmation and provisional adherence, the Joint Meeting of the Chemicals Group and Management Committee of the Special Programme on the Control of Chemicals (Joint Meeting) would organise, in consultation with the non-member country, technical support that might assist in the implementation of the Council Acts.

iv) The non-member country would be invited by the Joint Meeting to nominate a Test Guideline Co-ordinator and to take part in the activities and meetings related to the development and updating of OECD Test Guidelines and to take part in technical meetings related to GLP and, if recommended by the OECD Panel on GLP, to attend as an observer meetings of the Panel. Such an invitation would be for a maximum of three years and could be renewed by the Joint Meeting.

v) Once the non-member country has fully implemented the Council Acts, and taking account of the recommendation of the Joint Meeting in this respect, the non-member country may be invited by the Council to adhere to the Council Acts and to join the part of the OECD Chemicals Programme involving the mutual acceptance of data as a full member; this would require the non-member country to contribute to the resource costs of implementing this part of the Chemicals Programme.

vi) Participation may be terminated by either party upon one year advance notice. The Council may set any further terms and conditions to the invitation.

OECD PUBLICATIONS, 2, rue André-Pascal, 75775 PARIS CEDEX 16
PRINTED IN FRANCE
(97 2005 10 1 P) ISBN 92-64-01282-6 – No. 54291 2005